The Logica Yearbook
2013

The Logica Yearbook 2013

Edited by

Michal Dančák
and
Vít Punčochář

© Individual authors and College Publications 2014
All rights reserved.

ISBN 978-1-84890-144-5

College Publications
Scientific Director: Dov Gabbay
Managing Director: Jane Spurr

www.collegepublications.co.uk

Original cover design by Laraine Welch
Printed by Lightning Source, Milton Keynes, UK

All rights reserved. No part of this publication may be reproduced, stored in a retrieval system or transmitted in any form, or by any means, electronic, mechanical, photocopying, recording or otherwise without prior permission, in writing, from the publisher.

Preface

The book that you are holding in your hands is a further entry in a series of volumes that aspires to make some of the ideas presented at the annual international symposium Logica permanently accessible to both the conference participants and the wider public. The symposium, which took place at Hejnice Monastery in the Czech Republic from June 17 to June 21, 2013, brought together logicians from many countries. This volume contains a representative sample of the contributions made at the conference.

The Logica symposium is an event with a tradition which reaches back to the late nineteen eighties. Since that time, it has evolved into a respected conference which has a firm place in the annual schedule of the international community of logicians. Though the symposium is open to researchers of both a mathematical and a philosophical bent, its audience traditionally consists mostly of logicians with philosophical interests. The informal atmosphere provides a space for a stimulating exchange of ideas among logicians of all generations, including students. As the editors of this volume we are proud that we can contribute to the successful completion of the yearly symposium cycle by presenting this collection to you.

Last year's Logica was—as were all previous Logica symposia—organized by the Department of Logic of the Institute of Philosophy of the Czech Academy of Sciences. More than thirty lectures were presented during the conference, including those read by a distinguished list of invited speakers: Johan van Benthem, Michael Dunn, Volker Halbach, and Michiel van Lambalgen. As happens every year, the conference was enriched by a social programme that provided room for friendly debates concerning professional topics as well as for the starting of personal friendships. The proceedings that are traditionally published within one year of the conference unfortunately offer only a very limited record of the topics discussed and cannot hope to even partially convey its atmosphere. In spite of that, we hope that you will find this book to be worthy of your attention.

Both the Logica symposium and The Logica Yearbook are the result of a joint effort by many people to whom we would like to express our gratitude. We are, of course, very grateful to the Institute of Philosophy for all the important support that made the event possible. We express our thanks to the staff of Hejnice Monastery for their hospitality and friendly assistance. Special thanks from the organizers and also, we believe, from the guests go to the Bernard Family Brewery of Humpolec, which has traditionally sponsored the social programme of the symposium by providing three barrels of its excellent beer. We owe thanks to the Grant Agency of the Czech Republic, which provided significant support for the meeting and for the publishing of this volume with the funding of the grant project no. 13-21076S. We would like to express our gratitude to Petra Ivaničová, who is the key member of the organizing crew. We would also like to thank College Publications and its managing director, Jane Spurr, for their very pleasant cooperation during the preparation of this book. Last, but not least, we would like to thank all of the authors for their exemplary collaboration during the editorial process.

Prague, May 2014

Vít Punčochář and Michal Dančák

Contents

Massimiliano Carrara, Daniele Chiffi and Davide Sergio:
    Knowledge and Proof: A Multimodal Pragmatic Language   1

Marie Duží, Jakub Macek and Lukáš Vích:
    Procedural Isomorphism and Synonymy   15

Chris Fox:
    Curry-Typed Semantics in TPL   35

Renata de Freitas, Leandro Suguitani and Petrucio Viana:
    Relation Algebra throughout Galois Connections   49

Jens Ulrik Hansen:
    Modeling Truly Dynamic Epistemic Scenarios in PDEL   63

John T. Kearns:
    1st and 3rd Person Logical Theories   77

Manfred Kupffer:
    Weak Logic of Modal Metaframes   91

Paweł Łupkowski:
    Compliance and Pure Erotetic Implication   105

Johannes Marti and Riccardo Pinosio:
    Topological Semantics for Conditionals   115

Vittorio Morato:
  Modal Validity and Actuality 129

Jaroslav Peregrin:
  Logic and Reasoning 143

Mario Piazza and Gabriele Pulcini:
  Strange Case of Dr. Soundness and Mr. Consistency 161

Vít Punčochář:
  Intensionalisation of Logical Operators 173

Hans Rott:
  Three Floors for the Theory of Theory Change 187

Igor Sedlár:
  Relating Logics of Justifications and Evidence 207

Petr Švarný, Ondrej Majer and Michal Peliš:
  Erotetic Epistemic Logic in Private Communication Protocol 223

Giacomo Turbanti:
  Perspectives in the Interpretation of Defeasible Reasoning 239

Knowledge and Proof: A Multimodal Pragmatic Language

MASSIMILIANO CARRARA, DANIELE CHIFFI
AND DAVIDE SERGIO

Abstract: In this paper we present a modal extension $\mathcal{L}^P_{\Box,K}$ of the pragmatic language $\mathcal{L}^P$ introduced by Dalla Pozza and Garola (1995). The descriptive part of $\mathcal{L}^P_{\Box,K}$ is the *fusion* of two modal languages, $\mathcal{L}_\Box$ and $\mathcal{L}_K$, endowed with two independent boxes, $\Box$ and K, interpreted as "it is proved that" and "it is known that". We start introducing the language and discussing some of its relevant properties. We then propose a semantic and pragmatic validity, and the connection between these two notions.

Keywords: multimodalities, pragmatic modal language, fusion of modal languages

1 Introduction

The Knowability Paradox is a logical argument showing that if all truths are knowable in principle, then all truths are, in fact, known. Many strategies have been suggested in order to avoid the above paradoxical conclusion. A family of solutions revising the logic underneath—called *logical revision*—has been proposed.

Adopting this general strategy for solving the paradox, in a recent paper some of us (Carrara & Chiffi, 2014) have sketched a language $K\mathcal{L}P$ where alethic and epistemic modality are combined with the pragmatic language for assertions $\mathcal{L}^P$ (introduced by Dalla Pozza and Garola (1995)). The goal of that paper was to have a logic where the paradox disappears. The intuitive idea was to propose an anti-realistic solution considering some epistemic aspects associated with the notion of *assertion*.

Aim of this paper is to analytically present a pragmatic language $\mathcal{L}^P_{\Box,K}$ given by extending the descriptive part of the pragmatic language for assertions $\mathcal{L}^P$ with multimodalities where alethic and epistemic modalities are combined.[1]

---
[1] For an introduction to epistemic logics see (Hendricks & Symons, 2009). For epistemic modality and its combination with other modalities see (van Benthem, 2006).

2 An outline of $\mathcal{L}^P$

A pragmatic language is the (disjoint) union of two sets of formulas: radicals and sentences. The set of radicals is the descriptive part of the language, while the set of sentences is its pragmatic part.

Radicals represent *propositional contents* of sentences. Sentences express illocutionary acts. A sentence is either *elementary*, i.e., obtained by prefixing a sign of pragmatic mood to a radical, or *complex*, i.e., obtained from other sentences by means of logical-pragmatic connectives ($\sim$, $\cap$, $\cup$, $\supset$, $\equiv$; signs for *pragmatic negation, conjunction, disjunction, implication,* and *biconditional* respectively).

Radicals and sentences have different syntactic forms and are interpreted in different ways. A pragmatic interpretation of a pragmatic language consists of both a semantic interpretation of its descriptive part and a pragmatic interpretation of its pragmatic part. Radicals are interpreted semantically in terms of truth-values, every radical is either *truth* or *false*. Sentences are interpreted pragmatically in terms of justification-values, so that every sentence is either *justified* or *unjustified*.

In the case of the pragmatic language for assertions $\mathcal{L}^P$, the descriptive part $\mathcal{L}$ is identified with the language of classical propositional logic (the set of propositional formulas), and the set of sentences is a set of assertions. Elementary sentences are thus built up using only the sign of pragmatic mood of assertion, $\vdash$. So, for example, if α_1 and α_2 are propositional formulas, then $\vdash \alpha_1$ and $\vdash \alpha_2$ are elementary assertions, while $\vdash \alpha_1 \cap \vdash \alpha_2$ or $\vdash \alpha_1 \cup \vdash \alpha_2$ are complex assertions. Consequently, in $\mathcal{L}^P$, there are no assertions whose contents are *modal* propositions. In order to overcome such a limitation, we introduce $\mathcal{L}^P_{\Box,K}$, a pragmatic language for assertions on *modal* propositional contents. In particular, the descriptive part $\mathcal{L}_{\Box,K}$ of $\mathcal{L}^P_{\Box,K}$ is the *fusion* (Carnielli & Coniglio, 2011) $\mathcal{L}_\Box \oplus \mathcal{L}_K$ of two modal languages, $\mathcal{L}_\Box$ and $\mathcal{L}_K$, endowed with two independent boxes, $\Box$ and K, interpreted as "it is proved that"[2] and "it is known that" respectively.[3]

In the rest of the paper we proceed in the following manner. We first introduce (Section 3) the pragmatic language and its (semantic and pragmatic) interpretations, then we discuss some of its relevant properties. In Section 4, we propose a semantic and pragmatic validity, and the connection between

---

[2] Where proof has to be understood in its intuitive sense.

[3] The fusion $\mathcal{L}_1 \oplus \mathcal{L}_2$ of two modal languages, $\mathcal{L}_1$ and $\mathcal{L}_2$, endowed with two independent boxes, $\Box_1$ and $\Box_2$, is the *smallest* modal language generated by both boxes. Note also that the fusion of modal languages is commutative. Hence: $\mathcal{L}_1 \oplus \mathcal{L}_2 = \mathcal{L}_2 \oplus \mathcal{L}_1$.

these two notions given by the justification lemma. We conclude with some remarks on the relation between proof and knowledge.

3 $\mathcal{L}^P_{\Box,K}$ and its interpretations

The set of radical formulas and the set of assertive formulas are defined recursively through the following set of formation rules: respectively

$$\alpha := p \mid \top \mid \bot \mid \neg\alpha \mid \alpha_1 \wedge \alpha_2 \mid \alpha_1 \vee \alpha_2 \mid \alpha_1 \to \alpha_2 \mid \alpha_1 \leftrightarrow \alpha_2 \mid \Box\alpha \mid K\alpha$$

$$\delta := \vdash \alpha \mid \sim\delta \mid \delta_1 \cap \delta_2 \mid \delta_1 \cup \delta_2 \mid \delta_1 \supset \delta_2 \mid \delta_1 \equiv \delta_2$$

In order to get a pragmatic interpretation of $\mathcal{L}^P_{\Box,K}$, we have to semantically interpret $\mathcal{L}^P_{\Box,K}$. This amounts to an interpretation of its descriptive part $\mathcal{L}_{\Box,K}$.

As a matter of fact, the semantics of the fusion $\mathcal{L}_1 \oplus \mathcal{L}_2$ of two modal languages, $\mathcal{L}_1$ and $\mathcal{L}_2$, endowed with two independent boxes, $\Box_1$ and $\Box_2$, is given within the class of frames of the form $\langle W, R_1, R_2 \rangle$ where $\langle W, R_1 \rangle$ and $\langle W, R_2 \rangle$ are frames for $\Box_1$ and $\Box_2$ respectively. The axiomatic presentation through Hilbert calculus is obtained by merging the axioms and the inference rules of both logics. Moreover, *Bridge Principles* (BPs) can be added, i.e. axioms intended to logically connect the independent boxes, e.g. $\Box_1\alpha \to \Box_2\alpha$.[4]

We are assuming $\mathcal{L}^P_{\Box,K}$ to be the *fusion* $\mathcal{L}_\Box \oplus \mathcal{L}_K$ of $\mathcal{L}_\Box$ and $\mathcal{L}_K$ endowed with $\Box$ and K intuitively interpreted as "it is proved that" and "it is known that" respectively. Hence, it is somehow intuitive to consider relational structures of the form $\langle W, R_\Box, R_K \rangle \in \mathcal{C}$ where $\mathcal{C}$ is the class of frames with W a set of possible worlds, and $R_\Box, R_K \subseteq W \times W$ binary accessibility relations on W such that $R_\Box$ is reflexive and transitive, while R_K is reflexive, symmetric and transitive. In this way, $\Box$ is an $S4 - like$ alethic modality, and K is an $S5 - like$ epistemic modality.

In addition, we introduce the following *Bridge Principle* (BP):

(BP) $\Box\alpha \to \neg K\neg\alpha$

which can be intuitively read as "if it is the case that α is proved to be true, then it is not the case that α is known to be false". (BP) gives a logical

---

[4]BPs can be equivalent to conditions on the relations between accessibility relations (Carnielli & Coniglio, 2011).

connection between $\Box$ and K that turns out to be equivalent to the condition that $R_K \subseteq R_\Box$ (Carnielli, Pizzi, & Bueno-Soler, 2008). Namely, (BP) is valid (on an appropriate frame) if and only if $R_K \subseteq R_\Box$.

The idea behind (BP) can be made clearer if we consider its equivalent formulation in terms of conjunction:

(BP') $\neg(\Box\alpha \wedge K\neg\alpha)$

(BP') identifies the relation expressing a minimal condition holding between proof and knowledge according to our pre-theoretical insights. That is, there must be a logical incompatibility between the proof that α is true and the knowledge that α is false.

The semantic and the pragmatic interpretation of $\mathcal{L}^P_{\Box,K}$ are given through the following definitions.

Definition 1 (Semantic interpretation of $\mathcal{L}^P_{\Box,K}$) *Let $\mathcal{C}$ be the class of Kripke frames $F = \langle W, R_\Box, R_K \rangle$ such that W is a set of possible worlds, $R_\Box$, $R_K \subseteq W \times W$ are binary accessibility relations on W, $R_\Box$ is reflexive and transitive, R_K is reflexive, symmetric and transitive, and $R_K \subseteq R_\Box$.*

Let $\mathcal{V}_F$ be the class of valuations

$$v : \begin{cases} PROP \to & \wp(W) \\ p \mapsto & v(p) \subseteq W \end{cases}$$

on a frame $F \in \mathcal{C}$ where $PROP$ is the set of atomic propositional radicals.

Let $\mathcal{M} = \{M = \langle F, v \rangle \,|\, F \in \mathcal{C} \,\&\, v \in \mathcal{V}_F\}$ be the class of models on a frame F.

Let $M = \langle \langle W, R_\Box, R_K \rangle, v \rangle \in \mathcal{M}$.

Then, the semantic interpretation σ_v of $\mathcal{L}^P_{\Box,K}$ on M is the function

$$\sigma_v : \begin{cases} (\mathcal{L}_\Box \oplus \mathcal{L}_K) \times W \to \{T, F\} \\ (\alpha, w) \mapsto \sigma_v(\alpha, w) \in \{T, F\} \end{cases}$$

which satisfies the following truth-rules:

(TR1) Let $p \in PROP$ and $w \in W$. Then:

(i) $\sigma_v(\top, w) = T$

(ii) $\sigma_v(\bot, w) = F$

(iii) $\sigma_v(p, w) = T \Leftrightarrow p \in v(p)$

(TR2) Let $\alpha, \alpha_1, \alpha_2 \in \mathcal{L}_{\Box,K}$ and $w \in W$. Then:

(i) $\sigma_v(\neg \alpha, w) = T \Leftrightarrow \sigma_v(\alpha, w) = F$

(ii) $\sigma_v(\alpha_1 \wedge \alpha_2, w) = T \Leftrightarrow \sigma_v(\alpha_1, w) = T$ and $\sigma_v(\alpha_2, w) = T$

(iii) $\sigma_v(\alpha_1 \vee \alpha_2, w) = T \Leftrightarrow \sigma_v(\alpha_1, w) = T$ or $\sigma_v(\alpha_2, w) = T$

(iv) $\sigma_v(\alpha_1 \to \alpha_2, w) = T \Leftrightarrow \sigma_v(\alpha_2, w) = T$ whenever $\sigma_v(\alpha_1, w) = T$

(v) $\sigma_v(\alpha_1 \leftrightarrow \alpha_2, w) = T \Leftrightarrow \sigma_v(\alpha_1 \to \alpha_2, w) = T$
and $\sigma_v(\alpha_2 \to \alpha_1, w) = T$

(TR3) Let $\alpha \in \mathcal{L}_{\Box,K}$ and $w, v \in W$. Then:

(i) $\sigma_v(\Box \alpha, w) = T \Leftrightarrow \sigma_v(\alpha, v) = T$ whenever $wR_\Box v$

(ii) $\sigma_v(K\alpha, w) = T \Leftrightarrow \sigma_v(\alpha, v) = T$ whenever $wR_K v$

Definition 2 (Pragmatic Interpretation of $\mathcal{L}^P_{\Box,K}$) *Let σ_v be a semantic interpretation of $\mathcal{L}^P_{\Box,K}$ on a model M. Then a pragmatic interpretation π_{σ_v} of $\mathcal{L}^P_{\Box,K}$ on M is a (partial) function*

$$\pi_{\sigma_v} : \begin{cases} \mathcal{L}^P_{\Box,K} \times W \to \{J, U\} \\ (\delta, w) \mapsto \pi_{\sigma_v}(\delta, w) \in \{J, U\} \end{cases}$$

such that it satisfies the following Justification Rules (JRs) and the Correctness Criterion (CC):

(JR1) Let $\alpha \in \mathcal{L}_{\Box,K}$ and $w \in W$. Then:

$\pi_{\sigma_v}(\vdash \alpha, w) = J \Leftrightarrow$ a proof exists that $\sigma_v(\alpha, w) = T$

Hence, $\pi_{\sigma_v}(\vdash \alpha, w) = U \Leftrightarrow$ no proof exists that $\sigma_v(\alpha, w) = T$

(JR2) Let $\delta, \delta_1, \delta_2 \in \mathcal{L}^P_{\Box,K}$ and $w \in W$. Then:

(i) $\pi_{\sigma_v}(\sim \delta, w) = J \Leftrightarrow$ a proof exists that $\pi_{\sigma_v}(\delta, w) = U$

(ii) $\pi_{\sigma_v}(\delta_1 \cap \delta_2, w) = J \Leftrightarrow \pi_{\sigma_v}(\delta_1, w) = J$ and $\pi_{\sigma_v}(\delta_2, w) = J$

(iii) $\pi_{\sigma_v}(\delta_1 \cup \delta_2, w) = J \Leftrightarrow \pi_{\sigma_v}(\delta_1, w) = J$ or $\pi_{\sigma_v}(\delta_2, w) = J$

(iv) $\pi_{\sigma_v}(\delta_1 \supset \delta_2, w) = J \Leftrightarrow$ a proof exists that $\pi_{\sigma_v}(\delta_2, w) = J$ whenever $\pi_{\sigma_v}(\delta_1, w) = J$

(v) $\pi_{\sigma_v}(\delta_1 \equiv \delta_2, w) = J \Leftrightarrow \pi_{\sigma_v}(\delta_1 \supset \delta_2, w) = J$ and $\pi_{\sigma_v}(\delta_2 \supset \delta_1, w) = J$

(CC) Let $\alpha \in \mathcal{L}_{\Box,K}$ and $w \in W$. Then $\pi_{\sigma_v}(\vdash \alpha, w) = J \Rightarrow \sigma_v(\alpha, w) = T$

4 Semantic and pragmatic validity for $\mathcal{L}^P_{\Box,K}$

We consider now the semantic and pragmatic notions of validity for $\mathcal{L}^P_{\Box,K}$. Intuitively, a radical is semantically valid (*s*-valid) for $\mathcal{L}^P_{\Box,K}$ if it is a *classical logical law* for its descriptive part.

It is worth noting that by taking into account soundness and completeness results about modal logics (Blackburn, de Rijke, & Venema, 2001) and their fusions (Fine & Schurz, 1991; Gabbay & Shehtman, 1998)—given the fact that the axiomatic presentation through Hilbert calculus of the fusion of modal logics is obtained by merging the axioms and the inference rules of the fused logics, and considering our (BP)—it is possible to introduce definitions and to obtain results about *s*-validity into the following proposition.

Proposition 1 *(Semantic validity for $\mathcal{L}^P_{\Box,K}$)*
Let $\Lambda \subseteq \mathcal{L}_{\Box,K} := \mathcal{L}_\Box \oplus \mathcal{L}_K$ be the normal modal logic given by the following axioms and closed under the following rules:

(TAUT) all the schemas of propositional tautologies belong to Λ;

(AX) the following axiom schemas belong to Λ:

(1) $\Box(\alpha_1 \to \alpha_2) \to (\Box \alpha_1 \to \Box \alpha_2)$ (P-Distributivity)

(2) $\Box \alpha \to \alpha$ (Correctness)

(3) $\Box\alpha \to \Box\Box\alpha$

(4) $K(\alpha_1 \to \alpha_2) \to (K\alpha_1 \to K\alpha_2)$ (K-Distributivity)

(5) $K\alpha \to \alpha$ (Factivity)

(6) $K\alpha \to KK\alpha$ (Positive Introspection)

(7) $\neg K\alpha \to K\neg K\alpha$ (Negative Introspection);

(BP) $\Box\alpha \to \neg K\neg\alpha$ (Proof-Knowledge Compatibility);

(MP) If $\alpha_1 \in \Lambda$ and $\alpha_1 \to \alpha_2 \in \Lambda$, then $\alpha_2 \in \Lambda$ (Modus Ponens);

(N) If $\alpha \in \Lambda$, then $\Box\alpha, K\alpha \in \Lambda$ (Generalization);

(UF) Uniform substitution.

Let $\alpha \in \mathcal{L}_{\Box,K}$. Then

(SV) $\alpha \in \mathcal{L}_{\Box,K}$ is *s*-valid for $\mathcal{L}^P_{\Box,K} \Leftrightarrow \alpha \in \Lambda$

Similarly to the *s*-validity for radicals; since pragmatic sentences are interpreted in terms of justification-values, it makes sense to define a sentence as pragmatically valid (*p*-valid) if it is justified in any case. Therefore:

An assertion $\delta \in \mathcal{L}^P_{\Box,K}$ is *p*-valid if $\pi_{\sigma_v}(\delta, w) = J$ for every π_{σ_v}.

Moreover, it turns out that the two notions of validity are closely related to each other by the *Justification Lemma*. The lemma is based on a *modal translation* of the pragmatic part of $\mathcal{L}^P_{\Box,K}$ into its descriptive part $\mathcal{L}_{\Box,K}$, viz. a syntactic translation of any assertion into a *modal* radical formula. There are two intuitive motivations behind the translation. The first one is that an (elementary) assertion is justified just in case there is a proof, intuitively left unspecified, that what is asserted is true. The second reason is that the way proofs, and so justification-values, are combined into the (JRs) is captured by the Brouwer-Heyting-Kolmogorov (BHK) interpretation of intuitionistic logic (Troelstra & Schwichtenberg, 2000). Therefore, the justifications of assertions can be formalized by means of the Gödel-McKinsey-Tarski *modal* translation of intuitionistic logic in terms of an $S4 - like$ modality (Gödel, 1933). Statements like "a proof exists that α is true" can be translated with the formula $\Box\alpha$, and any complex assertion can be translated into a modal formula according to the modal translation of its BHK reading.

The following definition makes the translation precise.

Definition 3 (Modal Translation of $\mathcal{L}^P_{\square,K}$ into $\mathcal{L}_{\square,K}$) Let $\mathcal{L}^P_{\square,K}$ and let $(-)^*$ *be the function:*

$$(-)^* : \begin{cases} \mathcal{L}^P_{\square,K} \to \mathcal{L}_{\square,K} \\ \delta \mapsto (\delta)^* \end{cases}$$

such that:

(MT1) Let $\alpha \in \mathcal{L}_{\square,K}$. Then:

$(\vdash \alpha)^* = \square \alpha$

(MT2) Let $\delta, \delta_1, \delta_2 \in \mathcal{L}^P_{\square,K}$. Then:

(i) $(\sim \delta)^* = \square \neg (\delta)^*$

(ii) $(\delta_1 \cap \delta_2)^* = (\delta_1)^* \wedge (\delta_2)^*$

(iii) $(\delta_1 \cup \delta_2)^* = (\delta_1)^* \vee (\delta_2)^*$

(iv) $(\delta_1 \supset \delta_2)^* = \square((\delta_1)^* \to (\delta_2)^*)$

(v) $(\delta_1 \equiv \delta_2)^* = \square((\delta_1)^* \leftrightarrow (\delta_2)^*)$

Proposition 2 (*Justification Lemma for $\mathcal{L}^P_{\square,K}$*)

Let $\delta \in \mathcal{L}^P_{\square,K}$ and $(\delta)^*$ be its modal translation. Then, for every pragmatic interpretation π_{σ_v} of $\mathcal{L}^P_{\square,K}$ we have that:

(JL) $\pi_{\sigma_v}(\delta, w) = J \Leftrightarrow \sigma_v((\delta)^*, w) = T$

On the basis of the justification lemma, we get a criterion for *p*-validity. The idea is the following: if justification-values can be reduced to truth-values, then *p*-validity can be reduced to *s*-validity as well.

Proposition 3 (*Pragmatic validity for $\mathcal{L}^P_{\square,K}$*)

Let $\delta \in \mathcal{L}^P_{\square,K}$ and $(\delta)^*$ be its modal translation. Then:

(PV) δ is p-*valid for* $\mathcal{L}^P_{\square,K} \Leftrightarrow (\delta)^*$ is s-*valid for* $\mathcal{L}^P_{\square,K}$

Knowledge and Proof

Remark 1 *Let us show an application of (PV). Considering (BP), it is not too difficult to see that from $\Box \alpha \to \neg K \neg \alpha$ it is possible to derive $\Box(\Box \alpha \to \Box \neg K \neg \alpha)$, and that $\Box(\Box \alpha \to \Box \neg K \neg \alpha)$ is the modal translation of $\vdash \alpha \supset \vdash \neg K \neg \alpha$, i.e. $\Box(\Box \alpha \to \Box \neg K \neg \alpha) = (\vdash \alpha \supset \vdash \neg K \neg \alpha)^*$. It follows that $\vdash \alpha \supset \vdash \neg K \neg \alpha$ is p-valid, and it could be read as the pragmatic version of (BP).*

Remark 2 *In $\mathcal{L}^P_{\Box, K}$ two versions of* Modus Ponens *can be formulated in the following ways:*

(**PMP1**) *If $\vdash (\alpha_1 \to \alpha_2)$ and $\vdash \alpha_1$, then $\vdash \alpha_2$*

(**PMP2**) *If $\delta_1 \supset \delta_2$ and δ_1, then δ_2*

Remark 3 *Here it is a list of* p-valid formulas, Pragmatic Bridge Principles, *that explain the relations between the semantic logical operators and the pragmatic ones.*

(**PBP1**) $(\vdash \neg \alpha) \supset (\sim \vdash \alpha)$

(**PBP2**) $\vdash (\alpha_1 \land \alpha_2) \equiv (\vdash \alpha_1 \cap \vdash \alpha_2)$

(**PBP3**) $(\vdash \alpha_1 \cup \vdash \alpha_2) \supset \vdash (\alpha_1 \lor \alpha_2)$

(**PBP4**) $\vdash (\alpha_1 \to \alpha_2) \supset (\vdash \alpha_1 \supset \vdash \alpha_2)$

(**PBP5**) $\vdash (\alpha_1 \leftrightarrow \alpha_2) \supset (\vdash \alpha_1 \equiv \vdash \alpha_2)$

(**PBP6**) $\vdash \Box \alpha \equiv \vdash \alpha$

(**PBP7**) $\vdash K\alpha \supset \vdash \alpha$

(**PBP8**) $\vdash K\alpha \supset \vdash \Box \alpha$

(**PBP9**) $\vdash K\alpha \equiv \vdash KK\alpha$

(**PBP10**) $\vdash \alpha \supset \vdash \neg K \neg \alpha$

(**PBP11**) $\vdash \neg K\alpha \supset \vdash K \neg K\alpha$

Notice that $\vdash \alpha$, $\vdash \Box \alpha$, and $\vdash \Box\Box\alpha$ are *p*-equivalent assertions, that $\vdash K\alpha$ and $\vdash KK\alpha$ are *p*-equivalent as well, but neither $\vdash K\alpha$ nor $\vdash KK\alpha$ is *p*-equivalent to $\vdash \alpha$. Indeed, $\Box(\Box K\alpha \leftrightarrow \Box \alpha)$ is not *s*-valid since

$\Box(\Box K\alpha \to \Box\alpha)$ is *s*-valid, but $\Box(\Box\alpha \to \Box K\alpha)$ is not. Therefore, by (PV), (PBP7) $\vdash K\alpha \supset \vdash \alpha$ is *p*-valid, but $\vdash \alpha \supset \vdash K\alpha$ is *not* p-valid.

In other words, because factivity holds, a proof that α is known to be true, is *actually* transformed into a proof that α is true.

Remark 4 *Here is a list of some* p-*valid formulas that could be of some interest.*

(P1) $\delta_1 \supset (\delta_2 \supset \delta_1)$

(P2) $(\delta_1 \supset (\delta_2 \supset \delta_3)) \supset ((\delta_1 \supset \delta_2) \supset (\delta_1 \supset \delta_3))$

(P3) $(\delta_1 \cap \delta_2) \supset \delta_{1/2}$

(P4) $\delta_1 \supset (\delta_2 \supset (\delta_1 \cap \delta_2))$

(P5) $\delta_{1/2} \supset \delta_1 \cup \delta_2$

(P6) $(\delta_1 \supset \delta_3) \supset ((\delta_2 \supset \delta_3) \supset (\delta_1 \cup \delta_2 \supset \delta_3))$

(P7) $(\delta_1 \supset \delta_2) \supset (\delta_1 \supset \sim \delta_2) \supset \sim \delta_1$

(P8) $(\delta_1 \equiv \delta_2) \supset ((\delta_1 \supset \delta_2) \cap (\delta_2 \supset \delta_1))$

(P9) $(\delta_1 \supset \delta_2) \supset ((\delta_2 \supset \delta_1) \supset (\delta_1 \equiv \delta_2))$

5 Some philosophical remarks on $\mathcal{L}^P_{\Box,K}$

$\mathcal{L}^P$ is a *language for assertions* mainly inspired by Frege and Dummett and by Austin's theory of illocutionary acts. Roughly speaking, the idea is to follow Frege distinguishing propositions from judgments: A proposition is either true or false, while a judgment, that can be expressed through the speech act of an assertion[5], is—according to Dalla Pozza and Garola's view—either justified or unjustified. A justified assertion is defined in terms of the existence of a proof that the asserted content is true. Although the concept of proof is meant to be *intuitive* and *unspecified*, it must always be understood as *correct*: a proof is a proof of the truth. The key ideas behind the language are both the explication of the notion of assertion on a content in terms of justification-values, and the definition of justification-values in

---

[5]Notice that Frege's analysis is extendable to other speech acts such as asking, questioning, etc. So is $\mathcal{L}^P$. Languages where $\mathcal{L}^P$ is expanded so to give rise of other pragmatics acts have been studied. See, for example, (Bellin & Biasi, 2004).

Knowledge and Proof

terms of the existence of proof of the truth of the asserted content. Therefore, the existence of a proof *is* the ground for a (justified) assertion. In $\mathcal{L}^P$, however, only assertions on propositional contents are expressible, and assertions on modal propositional contents are not allowed. Here, we are interested in expanding the expressiveness of the assertions from propositional contents to modal (propositional) contents, and, in particular, to assertions on *alethic* and *epistemic* contents. For doing it, we extend the descriptive part of $\mathcal{L}^P$ introducing $\mathcal{L}^P_{\Box,K}$, a *multimodal* pragmatic language for assertions on modal contents. The assertable modal contents are given by combining (*fusion* of) two modal languages, $\mathcal{L}_\Box$ and $\mathcal{L}_K$, endowed with two independent boxes, $\Box$ for the alethic modality and K for the epistemic one. These two boxes are intuitively interpreted as "it is proved that" and "it is known that" and formally realized by means of an $S4 - like$ and an $S5 - like$ modality respectively.

Given the intuitive interpretation of the $\Box$, it is possible to read alethic contents intuitionistically. Indeed, via BHK interpretation, an intuitionistic "truth" is defined in terms of the existence of a proof. We interpret $\Box$ in this way for two reasons.

The first one is more a technical reason, and it is related to the fact that any pragmatic language for assertions is *essentially* an intuitionistic language: As we saw, the existence of a proof *is* the ground for a (justified) assertion. Therefore, if the descriptive part of a pragmatic language has an $S4 - like$ box, then its pragmatic part can be *projected* into its descriptive one by means of a syntactic translation, i.e., the function (-)*. (JRs) are reduced to truth-rules on the correspondent modal translations and the other way around (JL), and methods and results from modal logic can be applied to study formally pragmatic languages (PV). The second one is related to the paradox of knowability. Indeed, if *alethic* notions have an intuitionistic$-like$ semantics, then "it is possible that", the dual notion of "it is necessary that", can be interpreted as "there is no proof that not". In such a way, the possibility of something to be true is reduced to the (actual) absence of a proof of its falsity, and $\Diamond K\alpha$ becomes "(at this moment in time), there is no proof that $K\alpha$ is false".

In reference to a more general study of pragmatic languages, there are several pressing issues of different kind that should be addressed. First, there are some *ontological* issues. *What sort of "things" are proofs? Are proofs processes or can they be thought of as objects?* Is the existence of a proof to be meant as an *actual* existence or a *potential* one? Are proofs tenseless or tensed "things"? Secondly, there are *linguistic* questions. Within an epistemic framework, for example, one can ask whether or not it exists the (performative speech) *act of knowing* the proposition α, and if it exists, whether or not it is reducible to the act of asserting the epistemic content $K\alpha$, i.e., $\vdash K\alpha$. And there are, of course, *epistemological* problems.

In this paper we do not go into any details about such pressing issues. We only assume that proofs and knowledge are not the same "things" and that, according to our pre-theoretical insights, the notion of knowledge is subject-related and tensed in the sense that it is more like a process than like an object. So, according to our insights, we propose only a minimal condition holding between *proof* and *knowledge*: it is assumed that there is a logical incompatibility between the proof that α is true and the knowledge that α is false, i.e.:

(**BP**$'$)$\neg(\Box\alpha \wedge K\neg\alpha)$

or its equivalent formulation:

(**BP**)$(\Box\alpha \to \neg K\neg\alpha)$.

Other BPs connecting knowledge and proof may be considered. Some of them are intuitively invalid, and, for others, explicit philosophical assumptions and clarifications are needed. Take, for example:

$$(\Box\alpha \leftrightarrow K\alpha) \qquad (1)$$

and

$$(\Box\alpha \to K\alpha). \qquad (2)$$

According to (1), proof and knowledge are, at least formally, the same "thing". But, this does not reflect our intuitions on processes *vs.* objects.

According to (2), given the existence of a proof for α, we know that α. Now, one may argue that, it is easy to observe that the mere existence of a proof of α is not sufficient to warrant that α is known: There might be proofs of α that we ignore. However, such an argument presupposes a more complete and consistent position about some ontological matters. Indeed, two readings of the sentence "there is a proof of α" can be specified: a tenseless and a tensed reading. If one argues, according to the tensed interpretation, that "there is a proof of α" means that *as a matter of fact, we either have proved α or at some time we shall prove it*, then the relation between knowledge and proof is trivial. When it happens to be a proof of α at time t, we know α and *vice versa*. Such an argument would make (2) valid, but would also validate (1).

References

Blackburn, P., de Rijke, M., & Venema, Y. (2001). *Modal Logic*. Cambridge: Cambridge University Press.

Carnielli, W. A., & Coniglio, M. E. (2011). Combining Logics. In E. N. Zalta (Ed.), *The Stanford Encyclopedia of Philosophy* (Winter 2011 ed.).

Carnielli, W. A., Pizzi, C., & Bueno-Soler, J. (2008). *Modalities and Multimodalities*. Berlin, Heidelberg: Springer.

Carrara, M., & Chiffi, D. (2014). The Knowability Paradox in the Light of a Logic for Pragmatics. In R. Ciuni, H. Wansing, & C. Willkommen (Eds.), *Recent Trends in Philosophical Logic* (pp. 33–48). Berlin: Springer.

Dalla Pozza, C., & Garola, C. (1995). A Pragmatic Interpretation of Intuitionistic Propositional Logic. *Erkenntnis*, *43*, 81–109.

Fine, K., & Schurz, G. (1991). Transfer Theorems for Stratified Multimodal Logics. In B. J. Copeland (Ed.), *Logic and Reality* (pp. 169–213). Oxford: Clarendon Press.

Gabbay, D. M., & Shehtman, V. B. (1998). Products of Modal Logics, Part 1. *Logic Journal of IGPL*, *6*, 73–146.

Gödel, K. (1933). Eine Interpretation des Intuitionistischen Aussagenkalkuls. *Ergebnisse Eines Mathematischen Kolloquiums*, *4*, 39–40.

Hendricks, V., & Symons, J. (2009). Epistemic Logic. In E. N. Zalta (Ed.), *The Stanford Encyclopedia of Philosophy* (Spring 2009 ed.).

Troelstra, A. S., & Schwichtenberg, H. (2000). *Basic Proof Theory*. Cambridge: Cambridge University Press.

van Benthem, J. (2006). Epistemic Logic and Epistemology: The State of Their Affairs. *Philosophical Studies*, *128*, 49–76.

Massimiliano Carrara
University of Padua
Italy
E-mail: massimiliano.carrara@unipd.it

Daniele Chiffi
University of Padua
Italy
E-mail: daniele.chiffi@unipd.it

Davide Sergio
E-mail: sdavide73@gmail.com

Procedural Isomorphism and Synonymy

Marie Duží, Jakub Macek and Lukáš Vích[1]

Abstract: Expressions are synonymous iff they have the same meaning. This simple definition evokes many problems including, inter alia, questions like what is the meaning of an expression and how fine-grained meanings should be. In TIL, which is our background theory, the sense of an expression is an algorithmically *structured procedure* detailing what operations to apply to what procedural constituents to arrive at the object (if any) denoted by the expression. Such procedures are rigorously defined as TIL *constructions*. However, in this new orthodoxy of *procedural semantics* we encounter the problem of the granularity of the individuation of procedures, because from the procedural point of view TIL constructions are a bit too fine-grained. In an effort to solve the problem we introduced the notion of *procedural isomorphism*. Procedural isomorphism is a nod to Carnap's intensional isomorphism and Church's synonymous isomorphism. Any two terms or expressions whose respective meanings are procedurally isomorphic are deemed semantically indistinguishable, hence synonymous. Yet the problem of the granularity of meanings remains open, because we have got several alternatives of the definition of the relation of procedural isomorphism on the set of TIL constructions. This is a pressing issue, because in a hyperintensional context that is neither intensional nor extensional only expressions with procedurally isomorphic meanings can be mutually substituted. The novel contribution of this paper is the proposal of a new definition of procedural isomorphism, viz. isomorphism modulo α-, and β-convertibility *by value*. This definition is an adjustment of Church's Alternative (A1). We argue for β-conversion by value and show that this conversion rather than Church's λ-conversion is the right way of applying a function to its argument.

Keywords: procedural semantics, beta conversion by value, procedural isomorphism, Transparent Intensional Logic, synonymy

---

[1] This research has been supported by the internal grant agency of VSB-TU Ostrava, project No. SP2014/157, "Knowledge modelling, process simulation and design", and co-financed by the European Social Fund and the state budget of the Czech Republic, project No. CZ.1.07/2.2.00/28.0216 "Logika: systémový rámec rozvoje oboru v ČR a koncepce logických propedeutik pro mezioborová studia" (Logic: the Development of the Discipline and Basic Logic Courses).

1 Introduction

In possible-world semantics that was a prevailing semantic theory in the last century, meanings are possible-world intensions and they are identical if co-intensional. Co-intensionality is nothing other than necessary co-extensionality. Yet already Carnap (1947, §13) pointed out that there are attitudinal contexts that are neither extensional nor intensional. Thus the consequences of possible-world semantics are well-known; linguistic senses and attitude contents are too coarsely individuated; attitudes proliferate too easily, etc.

Since the late sixties of the last century many logicians have striven for *hyperintensional semantics* and *structured meanings*. Moschovakis (1994) comes with an idea of *meaning as algorithm*. Yet much earlier, Tichý (1968, 1969) formulated the idea of *procedural* (as opposed to set-theoretical denotational) *semantics*, according to which the sense of an expression is an algorithmically structured procedure detailing what operations to apply to what procedural constituents to arrive at the object (if any) denoted by the expression. Such procedures are rigorously defined as TIL *constructions*. Tichý developed a logical framework known today as *Transparent Intensional Logic* (TIL) that serves as our background theory in this paper. In modern jargon, TIL belongs to the paradigm of *structured meaning*. However, in the new orthodoxy of structured meaning we encounter two major outstanding issues. One is the granularity of the individuation of structures. The other is the unity of structures. In this paper we are going to deal with the former, because it is a pressing issue. Only expressions with 'structurally isomorphic' meanings are synonymous and can be mutually substituted in hyperintensional contexts.

The topic of hyperintensionality was born out of *negativity*, as it were. As mentioned above, Carnap noticed that there are contexts which are neither extensional nor intensional, because the substitution of logically equivalent expressions fails here. Cresswell defines any individuation as hyperintensional that is finer than logical/necessary/strict equivalence. Hyperintensionality became originally a matter of *blocking* unwanted and unwarranted inferences, by pointing out that the correct substituends are hyperintensions. Indeed, any hyperintensional logic and formal semantics worth its name must be able to block various invalid inferences. But there is the other side of the coin, which is the *positive* topic of which inferences should be *validated*. That is, how hyper is hyper? If there is one central question permeating hyperintensional logic and semantics then this is it.

Procedural Isomorphism and Synonymy

Our definition of hyperintensionality is positive rather than negative. Any context in which the meaning of an expression is *displayed* rather than *executed* is hyperintensional.[2] Moreover, our conception of structured meaning is *procedural*. Hyperintensions are abstract procedures rigorously defined as TIL constructions which are assigned to expressions as their context-invariant meanings. The semantics is tailored to the hardest hyperintensional contexts, and generalized from there to simpler intensional and extensional contexts. This entirely anti-contextual and compositional semantics is, to the best of our knowledge, the only one that deals with all kinds of context, whether extensional, intensional or hyperintensional, in a uniform way. The same extensional logical laws are valid invariably in all kinds of context. In particular, there is no reason why Leibniz's law of substitution of identicals, and the rule of existential generalisation were not valid. What differ according to the context are not the rules themselves but the types of objects on which these rules are applicable. In an extensional context they are values of the functions denoted by the respective expression; in an intensional context they are the denoted functions themselves, and finally in a hyperintensional context they are procedures, that is the meanings themselves. Due to its stratified ontology of entities organised in a ramified hierarchy of types, TIL is a logical framework within which such an extensional logic of hyperintensions has been introduced.[3]

The syntax of TIL is Church's (higher-order) typed λ-calculus the terms of which are procedurally interpreted, which means that they denote structured modes of presentation of functions (that is TIL constructions) rather than set-theoretic functions. Thus, lambda abstraction transforms into the molecular procedure of forming a function and application into the molecular procedure of applying a function to an argument. Yet the problem of the granularity of meanings/procedures remains open, because TIL constructions are a bit too fine-grained from the procedural point of view. The main issue here is this. Constructions that differ at most by using different bound variables of the same type differ so slightly that we wish to say that such constructions are from the procedural point of view identical procedures.

Church aimed at defining the degree to which meaning should be fine-grained, and he proposed several Alternatives. Senses are identical if the respective expressions are (A0) *'synonymously isomorphic'* or (A1) mutu-

---

[2] In (Duží, Jespersen, & Materna, 2010) we use the terms 'mentioned' vs. 'used' construction. Since these terms are used for the use-mention distinction of expressions in linguistics, here we vote for 'displayed' vs. 'executed', respectively. See also (Duží & Jespersen, n.d.).

[3] See, for instance, (Duží, 2012, 2013).

ally λ-convertible. (A0) is α-conversion and synonymies resting on meaning postulates that assign composed meaning to constants; (A1) is α- and β-conversion; Church also considered Alternative (A1') that is α-, β- and η-conversion, while Alternative (A2) that is logical equivalence was rejected as a too weak criterion.[4]

In an effort to solve the problem of the procedural identity we introduce the notion of *procedural isomorphism*. Procedural isomorphism is a nod to Carnap's intensional isomorphism and Church's synonymous isomorphism. Any two terms or expressions whose respective meanings are procedurally isomorphic are deemed semantically indistinguishable, hence synonymous, hence substitutable in any context. The novel contribution of this paper is an adjustment of Church's Alternative (A1). The adjustment consists in the new definition of β-conversion, to wit β-conversion by value.

The rest of the paper is organised as follows. Section 2 sets out the foundations of our background theory TIL. In section 3 we introduce and define procedural isomorphism, Alternative (A1"), that rests on β-conversion *by value*. We prove that the so defined relation is a strict equivalence on the set of constructions. Concluding remarks are in section 4.

2 Logical foundations

Constructions are the *key* entities of TIL. They are algorithmically structured procedures, of one or multiple constituent parts and they serve to explicate linguistic meanings. Importantly, the constituent parts of a construction C are its *executed* subconstructions rather than the product (if any) of C which is located beyond C. Just to be clear, constructions are not functions, nor are they formulae or otherwise linguistic entities. Below we first define *simple types of order 1*, then we define *constructions* together with the types of their products, and finally the *ramified hierarchy of types*.

Definition 1 (types of order 1) *Let B be a base, i.e. a collection of pairwise disjoint, non-empty sets. Then:*

(i) *Every member of B is an elementary type of order 1 over B.*
(ii) *Let α and $\beta_1,\ldots,\beta_m$ ($m > 0$) be types of order 1 over B. Then the collection $(\alpha\ \beta_1 \ldots \beta_m)$ of all m-ary partial mappings from $\beta_1 \times \ldots \times \beta_m$ into α is a functional type of order 1 over B.*
(iii) *Nothing else is a type of order 1 over B.*

---
[4]For details see (Anderson, 1998).

For the purposes of natural-language analysis, we are currently assuming the following base of *ground types*:

- o: the set of truth-values $\{\mathbf{T}, \mathbf{F}\}$;
- ι: the set of individuals (constant universe of discourse);
- τ: the set of real numbers (doubling as temporal continuum);
- ω: the set of logically possible worlds (logical space).

Definition 2 (construction)

(i) (Variable) *Let valuation v assign object o to variable x. Then x v-constructs the object o.*

(ii) (Trivialization) *Let X be any object whatsoever (i.e. an extension, an intension, or a construction). Then ${}^0X$ is the* Trivialization *of X, which constructs X without any change of X.*

(iii) (Composition) *Let X v-construct a function f of type $(\alpha\, \beta_1 \ldots \beta_m)$ and let $Y_1, \ldots, Y_m$ v-construct entities $B_1, \ldots, B_m$ of respective types $\beta_1, \ldots, \beta_m$. Then the* Composition $[X\, Y_1 \ldots Y_m]$ *v-constructs the value (an entity, if any, of type α) of f on the tuple argument $\langle B_1, \ldots, B_m \rangle$. Otherwise the* Composition $[X\, Y_1 \ldots Y_m]$ *does not v-construct anything and so is v-improper.*

(iv) (Closure) *Let $x_1, \ldots, x_m$ be pair-wise distinct variables v-constructing entities of types $\beta_1, \ldots, \beta_m$ and let Y be a construction v-constructing an α-entity. Then $[\lambda x_1 \ldots x_m\, Y]$ is the construction* λ-Closure *(or* Closure*). It v-constructs the following function f of type $(\alpha \beta_1 \ldots \beta_m)$. Let $v(B_1/x_1, \ldots, B_m/x_m)$ be a valuation identical with v at least up to assigning objects $B_1/\beta_1, \ldots, B_m/\beta_m$ to variables $x_1, \ldots, x_m$. If Y is $v(B_1/x_1, \ldots, B_m/x_m)$-improper (see (iii)), then f is undefined on $\langle B_1, \ldots, B_m \rangle$. Otherwise the value of f on $\langle B_1, \ldots, B_m \rangle$ is the α-entity $v(B_1/x_1, \ldots, B_m/x_m)$-constructed by Y.*

(v) (Single Execution) *Let X v-construct object o. Then the* Single Execution ${}^1X$ *v-constructs o. Let X be either a non-construction or a v-improper construction. Then ${}^1X$ is v-improper.*

(vi) (Double Execution) *Let X v-construct a construction Y and let Y v-construct object Z (possibly itself a construction). Then the* Double Execution ${}^2X$ *v-constructs Z. Otherwise ${}^2X$ is v-improper.*

(vii) *Nothing else is a* construction.

Here are some explicative remarks. A *variable* constructs an object by having that object as its value dependent on a valuation function v arranging variables and objects in a sequence. *Trivialization* is TIL's objectual counterpart of a non-descriptive constant term, which simply provides a particular object. *Composition* is the procedure of functional application, rather than the functional value (if any) resulting from application. *Closure* is the procedure of forming a function by lambda abstraction, rather than the resulting function. Variables and Trivializations are the *atomic* constructions of TIL, Composition and Closure are the *molecular* constructions. An *atomic* construction is a structured whole with but one constituent part, namely the construction itself. A *molecular* construction is a structured whole with more parts than just itself. Importantly, the only part of ^0X is ^0X and not X, which is located beyond ^0X: the product of a procedure is no part of the procedure.

The definition of the typed universe of TIL amounts to a definition of the ramified hierarchy of types which divides into three parts; firstly, simple types of order 1, which were already defined by Definition 1; secondly, constructions of order n; thirdly, types of order $n + 1$.

Definition 3 (ramified hierarchy of types)
T_1 *(types of order 1).* See Definition 1.
C_n (constructions of order n)

(i) *Let x be a variable ranging over a type of order n over B. Then x is a* construction of order n over B.

(ii) *Let X be a member of a type of order n over B. Then* ^0X, ^1X, ^2X *are* constructions of order n over B.

(iii) *Let $X, X_1, \ldots, X_m$ ($m > 0$) be constructions of order n over B. Then $[X X_1 \ldots X_m]$ is a* construction of order n over B.

(iv) *Let $x_1, \ldots, x_m, X$ ($m > 0$) be constructions of order n over B. Then $[\lambda x_1 \ldots x_m X]$ is a* construction of order n over B.

(v) *Nothing is a* construction of order n over B *unless it so follows from C_n (i)-(iv).*

T_{n+1} (types of order $n + 1$) *Let $*_n$ be the collection of all constructions of order n over B. Then*

(i) $*_n$ *and every type of order n are* types of order $n + 1$ *over B.*

(ii) *If $m > 0$ and $\alpha, \beta_1, \ldots, \beta_m$ are types of order $n + 1$ over B, then $(\alpha \beta_1 \ldots \beta_m)$ (see T_1 (ii)) is a* type of order $n + 1$ *over B.*

(iii) *Nothing else is a* type of order $n + 1$ *over B.*

Logical objects like *truth-functions* and *quantifiers* are extensional: ∧ (conjunction), ∨ (disjunction) and ⊃ (implication) are of type (*ooo*) and ¬ (negation) of type (oo). The *quantifiers* $\forall^\alpha, \exists^\alpha$ are type-theoretically polymorphous, total functions of type $(o(o\alpha))$, for an arbitrary type α, defined as follows. The *universal quantifier* $\forall^\alpha$ is a function that associates a class A of α-elements with **T** if A contains all elements of the type α, otherwise with **F**. The *existential quantifier* $\exists^\alpha$ is a function that associates a class A of α-elements with **T** if A is a non-empty class, otherwise with **F**. Below all type indications will be provided outside the formulae in order not to clutter the notation. Furthermore, 'X/α' means that an object X is (a member) of type α. '$X \to_v \alpha$' means that the type of the object v-constructed by X is α. We write '$X \to \alpha$' if what is v-constructed does not depend on a valuation v. This holds throughout: the variables $w \to_v \omega$ and $t \to_v \tau$. If $C \to_v \alpha_{\tau\omega}$ then the frequently used Composition $[[Cw]\,t]$, which is the intensional descent (a.k.a. extensionalization) of the α-intension v-constructed by C, will be encoded as 'C_{wt}'. When using constructions of truth-functions, we often omit Trivialization and use infix notation to conform to standard notation in the interest of better readability. Moreover, the outermost brackets of Closure will be occasionally omitted.

Definition 4 (subconstruction) *Let C be a construction. Then*

  (i) C *is a* subconstruction *of* C.
  (ii) *If C is* ^0X, ^1X *or* ^2X *and X is a construction then X is a subconstruction of C.*
  (iii) *If C is $[XX_1 \ldots X_n]$ then $X, X_1, \ldots, X_n$ are subconstructions of C.*
  (iv) *If C is $[\lambda x_1 \ldots x_n Y]$ then Y is a subconstruction of C.*
  (v) *If A is a subconstruction of B and B is a subconstruction of C then A is a subconstruction of C.*
  (vi) *A construction is a subconstruction of C only if it so follows from (i)–(v).*

There are two modes in which a subconstruction C of a construction D may occur, to wit *displayed* and *executed*. If the latter, then we say that C is a *constituent* of D. The Trivialization ^0C displays the construction C and all the subconstructions of C. Hence C is not a constituent part of ^0C; it is not executed, and so does not obtain an object beyond it. We say that C occurs hyperintensionally. It is however important to realise that Double Execution executes a construction twice over. Thus in $^{2 0}C$ the subconstruction C is a constituent part of $^{2 0}C$.

If C is an executed constituent of D then C can occur *intensionally* or *extensionally*. In principle, constituent C occurs in D intensionally if C

is not Composed with a construction of the argument of the function f, v-constructed by C. Hence the whole function f is an object of predication within D. On the other hand, a constituent C of D occurs extensionally if it is Composed with a construction of an argument of the function f. Hence the *value* (if any) of the function f is an object of predication within D.

These three ways in which a subconstruction C of a construction D can occur give rise to three kinds of context.[5]

- *hyperintensional context*: a construction C occurs *displayed* and serves itself as a functional argument (though a construction of a higher order needs to be executed in order to produce the displayed construction);
- *intensional context*: construction C occurs *executed* in order to produce a function f but not the value of f (moreover, the executed construction does not occur within another hyperintensional context). Hence the entire *function*, v-constructed by C serves as a *functional argument*;
- *extensional context*: construction C occurs *executed* in order to produce a particular value of the function, v-constructed by C (moreover, the executed construction does not occur within another intensional or hyperintensional context). Hence the *value* of the function, v-constructed by C serves as a *functional argument*;

Higher context is dominant over a lower one. It means that all the subconstructions of a displayed construction occur hyperintensionally as well, and all the subconstructions of an executed construction that occurs intensionally occur intensionally as well.

3 Procedural isomorphism

As mentioned above, our logic is an extensional logic of hyperintensions.[6] Hence the extensional rules of existential generalisation and the substitution of identicals must be applicable also in hyperintensional contexts. The rules of existential generalisation into intensional and hyperintensional contexts have been specified in (Duží & Jespersen, 2010) and adjusted in (Duží & Jespersen, n.d.). As for the substitution of identicals, in an extensional or intensional context there is no problem. In an *intensional* context analytically

---
[5]Here we present just a summary. For exact definitions see (Duží et al., 2010).
[6]In this section we partly draw on material from (Duží & Jespersen, n.d.).

equivalent constructions are substitutable, and in an *extensional* context also v-congruent constructions are substitutable. Constructions are *v-congruent* if they v-construct the same object for a given valuation v. Constructions are analytically *equivalent* if they v-construct the same object for every valuation v. Obviously, analytically equivalent constructions are v-congruent, but not vice versa.

However, substitution of merely analytically equivalent constructions is not valid in hyperintensional contexts, as already Carnap (1947, §13) in effect pointed out. From a linguistic point of view, in a hyperintensional context only synonymous expressions can be substituted, because the very meaning of expressions is displayed. Our thesis is that synonymous expressions have structurally isomorphic meanings. And since meaning is a procedure, we need to define the relation of *procedural isomorphism* between constructions, because constructions are a bit too fine-grained from the procedural point of view. The main issue here is this. Constructions that differ at most by using different λ-bound variables of the same type differ so slightly that we wish to say that such constructions are one and the same procedure. For instance, the Closures $\lambda x\ [^0+\ x\ ^01]$, $\lambda y\ [^0+\ y\ ^01]$, $\lambda z\ [^0+\ z\ ^01], \ldots$, are by Definition 2 different constructions of the successor function. Yet from the procedural point of view they are isomorphic. They consist of these steps:

- take the function plus (by $^0+$)
- take *any* number x (or y, or $z, \ldots$)
- take the number 1 (by ^{01})
- apply the function plus on the pair of the chosen number x (or y, or $z, \ldots$) and 1
- abstract over the chosen number (λx, or λy, or $\lambda z, \ldots$)

Church proposed several Alternatives to specify the criterion of synonymy. The weakest Alternative (A2) that is logical equivalence has been rejected by Church as being too permissive. (A1) includes α- and β-conversion, while the strongest (A0) includes α-conversion and meaning postulates for atomic constants such as 'bachelor' and 'fortnight'. Church's (A0) and (A1) leave room for additional Alternatives. One such would be (A$\frac{1}{2}$), another (A$\frac{3}{4}$). The former includes α- and η-conversion while the latter adds to these two *restricted* β-conversion *by name*. In (Duží et al., 2010, Chapter 2) we advocate for (A$\frac{1}{2}$) whereas in (Duží & Jespersen, 2012) we prefer (A$\frac{3}{4}$) to soak up those differences between β-transformations that concern only λ-bound variables and thus (at least appear to) lack natural-language counterparts.[7] The *restricted* version of *equivalent* β-reduction by name consists in substituting free variables for λ-bound variables of the same type. For instance, the Composition $[\lambda x\ [^0+\ x\ ^01]\ y]$ can be simplified to the Composition $[^0+\ y\ ^01]$. Thus this transformation is just a formal manipulation with λ-bound variables that has much in common with η- and less with β-reduction. The latter is the operation of applying a function f to its argument a in order to obtain the value of f at a (leaving it open whether a value emerges). No such features can be found in restricted β-reduction. It is just a formal simplification of the original construction.

Recently, however, we have grown discontent with both (A$\frac{1}{2}$) and (A$\frac{3}{4}$), and we suggest a new definition of procedural isomorphism, (A1"). This variant (A1") is very close to Church's (A1) and includes α- and β-conversion *by value*. Thus we exclude η-conversion, and introduce a new version of β-conversion.

There are two reasons for not including η-conversion. First, it is actually rather peculiar to claim that two procedures are isomorphic if they do not have the same number of constituents. Yet the η-expanded construction of the form $\lambda x\ [F\ x]$ has two more constituents than the equivalent η-reduced construction F, because the former adds the steps of applying the function v-constructed by F to the value v-constructed by the variable x followed by abstraction over the value of x. The second and more important reason is the fact that η-conversion does *not preserve logical equivalence* in a logic of *partial functions* such as TIL. To see this, consider the following example. Let $F \to ((\alpha\beta)\gamma)$ v-construct a function that is not defined at the argument v-constructed by $A \to \gamma$. Then the Composition $[F\ A] \to (\alpha\beta)$ is v-improper. However, the η-expanded construction $\lambda x\ [[F\ A]\ x] \to (\alpha\beta)$, $x \to \beta$, v-constructs a *degenerate function*, which is a function undefined at all its arguments. To be sure, due to the v-improperness of $[F\ A]$ the Composition $[[F\ A]\ x]$ is also v-improper. But λ-abstraction raises the context to an intensional one, hence the Closure $\lambda x\ [[F\ A]\ x]$ v-constructs a degenerate function, which *is* an object, if a bizarre one. Hence the constructions $[F\ A]$ and $\lambda x\ [[F\ A]\ x]$ are not logically equivalent.[8]

In practice the exclusion of η-conversion from the definition of procedural isomorphism is going to be harmless. When analyzing expressions in

---

[7]The first version of procedural isomorphism in TIL was Materna's *Quid-identity* in (Materna, 1998, §5.3).

[8]We are grateful to Jiří Raclavský for calling our attention to this problem. See also (Raclavský, 2010).

TIL we apply our method of *literal analysis*, which consists of three steps: (i) assigning types to the objects mentioned by the sub-terms of the analyzed expression E; (ii) combining the Trivializations of the objects mentioned by the semantically simple sub-terms of E in order to obtain the construction of the object (if any) denoted by E; (iii) checking whether the resulting construction is type-theoretically coherent. Due to step (ii) the application of this method yields a construction (namely the meaning of E) that does not contain η-expanded subconstructions. For instance, the literal analysis of "The Pope is wise" is the Closure $\lambda w \lambda t\ [^0Wise_{wt}\ {}^0Pope_{wt}]$ rather than $\lambda w \lambda t\ [\lambda w \lambda t\ [\lambda x\ [^0Wise_{wt}\ x]]_{wt}\ {}^0Pope_{wt}]$, because the literal analysis of the predicate 'is wise' is the Trivialization ^0Wise rather than the Closure $\lambda w \lambda t\ [\lambda x\ [^0Wise_{wt}\ x]]$. The types are $Wise/(o\iota)_{\tau\omega}; Pope/\iota_{\tau\omega}; x \to \iota$.

The reasons for excluding unrestricted β-conversion are these. Though it is the fundamental computational rule of the λ-calculi, it is underspecified by the commonly acknowledged rule $[\lambda x\ C(x)\ A] \vdash C(A/x)$.[9] The procedure of applying the function v-constructed by $\lambda x\ C(x)$ to the argument v-constructed by A can be executed in two different ways: *by value* or *by name*. If by name then the *procedure* A is substituted for all the occurrences of x into C. In this case there are two problems.

First, conversion by name is not guaranteed to be a logically equivalent transformation as soon as partial functions are involved. This is due to the fact that A occurs extensionally as a constituent of the left-hand side construction, whereas when dragged into C its occurrence may become intensional provided the context in which x occurs in C is intensional. For instance, the Composition $[\lambda x\ [\lambda y\ [^0+\ x\ y]]\ [^0{:}\ ^05\ ^00]]$ is improper, because $[^0{:}\ ^05\ ^00]$ is improper. This is as it should be, for there is no value that might be substituted for the formal parameter x. However, the β-reduced construction $[\lambda y\ [^0+\ [^0{:}\ ^05\ ^00]\ y]]$ is *not* improper as it constructs a degenerated function undefined at all its arguments. The improper construction $[^0{:}\ ^05\ ^00]$ has been drawn into the intensional context of the Closure $[\lambda y\ [^0+\ x\ y]]$.

Second, even in those cases when β-reduction is an equivalent transformation, it may yield a loss of analytic information on which function has been applied to which argument.[10] For instance, the Composition $\lambda w \lambda t\ [[\lambda x\ [^0Larger_{wt}\ x\ x]]\ a]$ which is the meaning of "a is larger than itself" reduces to $\lambda w \lambda t\ [^0Larger_{wt}\ a\ a]$, the meaning of "$a$ is larger than a". Yet the two sentences are not strictly synonymous, because in the former

---

[9]For the sake of simplicity, we now consider a unary function. Generalisation for n-ary functions is obvious.

[10]For details see (Duží & Jespersen, 2010, 2013).

the property of being larger than itself is applied to a while in the latter the binary relation larger than is applied to the pair (a, a).[11]

The idea of conversion by value is simple. Execute the procedure A first, and only if A does not fail to produce an argument value on which C is to operate, substitute (the Trivialization of) this *value* for x. The solution preserves logical equivalence, avoids the problem of the loss of analytic information, and moreover, in practice it is more efficient. The efficiency is guaranteed by the fact that the procedure A is executed only once, whereas if this procedure is substituted for all the occurrences of the λ-bound variable it can subsequently be executed more than once. Thus we define function $Sub^n/(*_n *_n *_n *_n)$:

Definition 5 (Sub^n) *Let constructions C_1, C_2, and C_3 v-construct constructions (of order n) D_1, D_2, and D_3, respectively. Then the Composition $[^0Sub^n\ C_1\ C_2\ C_3]$ v-constructs the construction D that results from D_3 by collisionless substitution of D_1 for all occurrences of D_2 in D_3.*

Occasionally we need the polymorphic function Tr^α defined as follows:

Definition 6 (Tr^α) *The function $Tr^\alpha/(*_n\ \alpha)$ returns as its value the Trivialization of its α-argument.*

For instance, let variable y v-construct entities of type ι, such as a. Then $[^0Tr^\iota\ y]\ v(a/y)$-constructs ^0a. This means that the Composition $[^0Sub^1\ [^0Tr^\iota\ y]\ ^0x\ ^0[^0F_{wt}\ x]]\ v(a/y)$-constructs the Composition $[^0F_{wt}\ ^0a]$. Note that there is a substantial difference between the *construction* Trivialization and the *function* Tr^α. Whereas ^0y constructs just the variable y regardless of valuation, y being ^{0}-bound in ^0y, $[^0Tr^\iota\ y]$ v-constructs the Trivialization of the object v-constructed by y. Hence y occurs free in $[^0Tr^\iota\ y]$. In what follows we simply write 'Sub' and 'Tr' omitting thus the type-superscripts whenever no confusion arises.

Definition 7 (β-conversion by value and β-equivalence) *Let $Y \to_v \alpha$; $x_1, D_1 \to_v \beta_1, \ldots, x_n, D_n \to_v \beta_n$, $[\lambda x_1 \ldots x_n Y] \to_v (\alpha \beta_1 \ldots \beta_n)$. Let C, D be constructions of the forms $[[\lambda x_1 \ldots x_n Y]D_1 \ldots D_n]$, $^2[^0Sub\ [^0Tr\ D_1]\ ^0x_1\ \ldots\ [^0Sub\ [^0Tr\ D_n]\ ^0x_n\ ^0Y]]$, respectively. Then the conversion $C \Rightarrow_\beta D$ is β-reduction by value. The reverse conversion is β-expansion by value. We will say that constructions C and D are β-equivalent.*

---

[11] See (Salmon, 2010).

Proposition 1 *Let C, D be β-equivalent constructions. Then C and D are strictly equivalent in the sense that for any valuation v they either v-construct one and the same entity or are both v-improper.*

Proof. Let C be of the form $[[\lambda x_1 \ldots x_n \, Y] \, D_1 \ldots D_n]$ and D of the form ${}^2[{}^0Sub \, [{}^0Tr \, D_1] \, {}^0x_1 \ldots [{}^0Sub \, [{}^0Tr \, D_n] \, {}^0x_n \, {}^0Y]]$. If some of the constructions $D_1, \ldots, D_n$ are v-improper then so are both C and D, according to Definition 2, (iii) and (vi). Otherwise, let $D_1, \ldots, D_n$ all be v-proper, v-constructing the objects $d_1, \ldots, d_n$, respectively. Then by Definition 2, (iv) the Closure $[\lambda x_1 \ldots x_n \, Y]$ v-constructs the following function f. If Y is $v(d_1/x_1, \ldots, d_n/x_n)$-improper, then f is undefined on $\langle d_1, \ldots, d_n \rangle$ and thus C is $v(d_1/x_1, \ldots, d_n/x_n)$-improper by Definition 2, (iii). Otherwise the value of f on $\langle d_1, \ldots, d_n \rangle$ is the α-entity $v(d_1/x_1, \ldots, d_n/x_n)$-constructed by Y. Let the entity $v(d_1/x_1, \ldots, d_n/x_n)$-constructed by Y be a. Then by Definition 2, (iii) of Composition, the construction C v-constructs a. We are to show that the construction D also v-constructs a. The first Execution of D constructs $Y(x_1/{}^0d_1, \ldots, x_n/{}^0d_n)$, i.e. the construction Y where according to the definition of the functions Sub and Tr all the occurrences of variables $x_1, \ldots, x_n$ are replaced by ${}^0d_1, \ldots, {}^0d_n$ respectively. Since the Trivializations ${}^0d_1, \ldots, {}^0d_n$ construct the entities $d_1, \ldots, d_n$, respectively, the second Execution $v(d_1/x_1, \ldots, d_n/x_n)$-constructs the entity a, or else nothing in case Y is $v(d_1/x_1, \ldots, d_n/x_n)$-improper. Hence C and D come out strictly equivalent. $\square$

β-conversion by value is the correct procedure of applying the function constructed by the Closure $\lambda x \, [F \, x]$ to the argument constructed by A. Unlike β-conversion by name, it is a strictly equivalent conversion that does not yield loss of analytic information. The Composition $[\lambda x \, [F \, x] \, A]$ is the specification of calling the procedure $\lambda x \, [F \, x]$ with the formal parameter x at the argument provided by the procedure A. It consists of these execution steps:

- execute A in order to obtain the argument value a; if A fails to v-construct anything (is v-improper) then abort the execution; else:
- take the Trivialization of the argument value a
- substitute the Trivialization of a for all the occurrences of the variable x in the procedure body F, and finally:
- execute the result of the substitution.

The Composition ${}^2[{}^0Sub \, [{}^0Tr \, A] \, {}^0x \, {}^0F]$ has exactly the same constituents. These are:

- A: execute A in order to obtain the argument value a; if A is v-improper then the entire Composition is v-improper; else:
- $[^0Tr\ A]$: obtain the Trivialization of a
- $[^0Sub\ [^0Tr\ A]\ ^0x\ ^0F]$: substitute the Trivialization of a for x in F
- $^2[^0Sub\ [^0Tr\ A]\ ^0x\ ^0F]$: execute the result.

We can see that β-reduction by value is the explicit specification of the procedure of applying the function constructed by $\lambda x\ [F\ x]$ to the argument constructed by A. Hence the term '$[\lambda x\ [F\ x]\ A]$' can be taken as an abbreviation of the full-fledged application specification given by '$^2[^0Sub\ [^0Tr\ A]\ ^0x\ ^0F]$'. For this reason it is reasonable to consider the two terms '$[\lambda x\ [F\ x]\ A]$' and '$^2[^0Sub\ [^0Tr\ A]\ ^0x\ ^0F]$' as being synonymous and thus substitutable in all contexts, including hyperintensional ones.

In order to define *procedural isomorphism* on the set of constructions, we still need another definition, to wit the definition of α-conversion. The standard definition that defines α-equivalent constructions as those that differ at most by using different λ-bound variables does not do, because the β-reduced constructions C, D which arise from α-equivalent constructions do not differ at most by using different λ-bound variables. For instance, constructions

$$[\lambda x\ [^0+\ x\ ^01]\ ^05]\ \text{and}\ [\lambda y\ [^0+\ y\ ^01]\ ^05]$$

are α-equivalent according to standard definition. Yet their β-reduced forms

$$^2[^0Sub\ [^0Tr\ ^05]\ ^0x\ ^0[^0+\ x\ ^01]]\ \text{and}\ ^2[^0Sub\ [^0Tr\ ^05]\ ^0y\ ^0[^0+\ y\ ^01]]$$

would not be α-equivalent. But they should be, because from the procedural point of view it is irrelevant which variables are used as formal parameters of the respective procedure. Thus we define:

Definition 8 (α-conversion) *Let C, D be constructions. Then C, D are α-equivalent, if either C, D differ at most by using different λ-bound variables, or their β-expanded forms differ at most by using different λ-bound variables.*

Proposition 2 *α-equivalent constructions are strictly equivalent by being either v-improper or v-constructing one and the same entity.*

Proof. As a consequence of Proposition 1 it is sufficient to prove that Closures of the form $[\lambda x_1 \ldots x_n\ Y(x_1, \ldots, x_n)]$, $[\lambda y_1 \ldots y_n\ Y(y_1, \ldots, y_n)]$, where $Y(x_1, \ldots, x_n)$ differs from $Y(y_1, \ldots, y_n)$ only by collisionless substitution of variables $x_1, \ldots, x_n$ for $y_1, \ldots, y_n$, respectively, v-construct one and the same function. But this immediately follows from Definition 2, (iv). □

Definition 9 (procedural isomorphism) *Let C, D be constructions. Then C, D are* procedurally isomorphic *iff either C and D are identical or there are constructions $C_1, \ldots, C_n$ ($n > 1$) such that $^0C = {^0C_1}$, $^0D = {^0C_n}$, and for each C_i, C_{i+1} ($1 \leq i < n$) it holds that C_i, C_{i+1} are either α- or β-equivalent.*

Corollary 1 *Procedural isomorphism is an equivalence relation defined on a set of constructions such that procedurally isomorphic constructions are strictly equivalent in the sense that for any valuation v they either v-construct one and the same entity or they are v-improper.*

Proof. Follows immediately from Propositions 1 and 2. □

Example 1 Let $\approx_\alpha / (o*_n*_n)$ and $\approx_\beta / (o*_n*_n)$ be the relations of α and β-equivalence respectively. That constructions C, D are α- or β-equivalent will be denoted by '$^0C \approx_\alpha {^0D}$', '$^0C \approx_\beta {^0D}$' respectively. Then the above constructions

$$[\lambda x\ [^0{+}\ x\ ^01]\ ^05],\ ^2[^0Sub\ [^0Tr\ ^05]\ ^0x\ ^0[^0{+}\ x\ ^01]]$$
$$[\lambda y\ [^0{+}\ y\ ^01]\ ^05],\ ^2[^0Sub\ [^0Tr\ ^05]\ ^0y\ ^0[^0{+}\ y\ ^01]]$$

are procedurally isomorphic, because the following equivalences hold:

$$^0[\lambda x\ [^0{+}\ x\ ^01]\ ^05] \approx_\alpha {^0[\lambda y\ [^0{+}\ y\ ^01]\ ^05]}$$
$$^0[\lambda x\ [^0{+}\ x\ ^01]\ ^05] \approx_\beta {^{02}[^0Sub\ [^0Tr\ ^05]\ ^0x\ ^0[^0{+}\ x\ ^01]]}$$
$$^0[\lambda y\ [^0{+}\ y\ ^01]\ ^05] \approx_\beta {^{02}[^0Sub\ [^0Tr\ ^05]\ ^0y\ ^0[^0{+}\ y\ ^01]]}$$
$$^{02}[^0Sub\ [^0Tr\ ^05]\ ^0x\ ^0[^0{+}\ x\ ^01]] \approx_\alpha {^{02}[^0Sub\ [^0Tr\ ^05]\ ^0y\ ^0[^0{+}\ y\ ^01]]}$$

Having defined procedural isomorphism, we can now specify the *rule of substitution in hyperintensional contexts*: In a hyperintensional context only procedurally isomorphic constructions are mutually substitutable.

Example 2 Consider the analysis of the sentence "Tilman is calculating the value of the function $sin(x){:}cos(x)$ at the argument equal to $\sqrt{2}$" that comes down to

$$\lambda w\ \lambda t\ [^0Cal_{wt}\ {}^0Tilman\ ^0[\lambda x\ [^0{:}\ [^0Sin\ x]\ [^0Cos\ x]]\ [^0\sqrt{}\ {}^02]]] \qquad (1)$$

Types: $Cal/(o\iota*_n)_{\tau\omega}$; $Tilman/\iota$; $:/(\tau\tau\tau)$; Sin, Cos, $\sqrt{}/(\tau\tau)$.

Since the construction $[\lambda x\ [^0{:}\ [^0Sin\ x]\ [^0Cos\ x]]\ [^0\sqrt{}\ {}^02]]$ is procedurally isomorphic with $^2[^0Sub\ [^0Tr\ [^0\sqrt{}\ {}^02]]\ ^0x\ ^0[^0{:}\ [^0Sin\ x]\ [^0Cos\ x]]]$, its β-reduced form, it follows that Tilman calculates the latter Composition as well as its α-equivalent procedurally isomorphic variant:

$$\lambda w\ \lambda t\ [^0Cal_{wt}\ {}^0Tilman\ ^{02}[^0Sub\ [^0Tr\ [^0\sqrt{}\ {}^02]]\ ^0x\ ^0[^0{:}\ [^0Sin\ x]\ [^0Cos\ x]]]] \qquad (2)$$

$$\lambda w\ \lambda t\ [^0Cal_{wt}\ {}^0Tilman\ ^{02}[^0Sub\ [^0Tr\ [^0\sqrt{}\ {}^02]]\ ^0y\ ^0[^0{:}\ [^0Sin\ y]\ [^0Cos\ y]]]]$$

But it does *not* follow that Tilman calculates the ratio of Sine at $\sqrt{2}$ and Cosine at $\sqrt{2}$:

$$\lambda w\ \lambda t\ [^0Cal_{wt}\ {}^0Tilman\ ^0[^0{:}\ [^0Sin\ [^0\sqrt{}\ {}^02]]\ [^0Cos\ [^0\sqrt{}\ {}^02]]]]$$

Note that $[^0{:}\ [^0Sin\ [^0\sqrt{}\ {}^02]]\ [^0Cos\ [^0\sqrt{}\ {}^02]]]$ is the result of β-reduction by name of the original Composition $[\lambda x\ [^0{:}\ [^0Sin\ x]\ [^0Cos\ x]]\ [^0\sqrt{}\ {}^02]]$. While in the latter the square root of 2 is calculated only once, in the reduced construction it is calculated twice. Hence the two constructions are not procedurally isomorphic.

Since *Cal* is a relation(-in-intension) of an individual to a *construction*, and the application of *Sub* produces a construction, a question arises here. Could we omit in (2) the Trivialisation and Double execution preceding the application of the function *Sub*? In other words, is the construction (3) equivalent to the above construction (2)?

$$\lambda w\ \lambda t\ [^0Cal_{wt}\ {}^0Tilman\ [^0Sub\ [^0Tr\ [^0\sqrt{}\ {}^02]]\ ^0x\ ^0[^0{:}\ [^0Sin\ x]\ [^0Cos\ x]]]] \qquad (3)$$

Our answer is *no*, it is not. The reason is this: in (3) Tilman is related to the *product* of $[^0Sub\ [^0Tr\ [^0\sqrt{}\ {}^02]]\ ^0x\ ^0[^0{:}\ [^0Sin\ x]\ [^0Cos\ x]]]$, namely to $[^0{:}\ [^0Sin\ {}^01.4142135\ldots]\ [^0Cos\ {}^01.4142135\ldots]]$. Obviously, (3) follows from (2) but not vice versa.

Yet there are other interesting issues concerning the interplay between Trivialisation and Double Execution. While Trivialisation raises the context up to the hyperintensional level, Double Execution decreases it back to the intensional one. Hence for any construction C it holds that $^2{}^0C$ is logically equivalent to C. The question is whether the two constructions, $^2{}^0C$ and C, are not procedurally isomorphic as well. In our opinion they are *not*. The former contains two additional executive steps, to wit Double Execution and Trivialisation. Though in an ordinary vernacular this slight difference would most probably not matter, in the semantics of a programming language it does matter.

Thus we are considering whether it is philosophically wise to adopt several notions of procedural isomorphism. The definition we proposed in this paper is a very strict criterion of synonymy, and it is not improbable that several degrees of hyperintensional individuation are called for, depending on which sort of discourse happens to be analysed. Thus we admit that slightly different definitions of procedural isomorphism are still thinkable. What appears to be synonymous in an ordinary vernacular might not be synonymous in a professional language like the language of, for instance, logic, mathematics or physics.

4 Conclusion

We demonstrated above how to validly apply Leibniz's Law of the substitution of identicals in hyperintensional contexts. Since in such a context the meaning of an expression is displayed, only synonymous expressions with the same meaning can be mutually substituted. Our criterion of synonymy is procedural isomorphism of the constructions expressed by the respective expressions. The paper offers a formally worked-out and philosophically motivated criterion of hyperintensional individuation, which is the relation of procedural isomorphism. The definition of procedural isomorphism includes a slightly more carefully stated version of α-conversion and β-conversion by value, which amounts to a modification of Church's Alternative (A1).

References

Anderson, C. A. (1998). Alonzo Church's Contributions to Philosophy and Intensional Logic. *The Bulletin of Symbolic Logic*, *4*, 129–171.

Carnap, R. (1947). *Meaning and Necessity: A Study in Semantics and Modal Logic*. Chicago: University of Chicago Press.

Duží, M. (2012). Towards an Extensional Calculus of Hyperintensions. *Organon F*, *19*, supplementary issue *1*, 20–45.

Duží, M. (2013). Deduction in TIL: From Simple to Ramified Hierarchy of Types. *Organon F*, *20*, supplementary issue *2*, 5–36.

Duží, M., & Jespersen, B. (n.d.). *Transparent Quantification into Hyperintensional Objectual Attitudes*. (submitted for publication)

Duží, M., & Jespersen, B. (2010). Transparent Quantication into Hyperintensional Contexts. In M. Peliš & V. Punčochář (Eds.), *The Logica Yearbook 2010* (pp. 81–98). London: College Publications.

Duží, M., & Jespersen, B. (2012). Transparent Quantification into Hyperintensional Contexts *de re*. *Logique & Analyse*, *220*, 513–554.

Duží, M., & Jespersen, B. (2013). Procedural Isomorphism, Analytic Information, and β-conversion by Value. *Logic Journal of the IGPL*, *21*, 291–308.

Duží, M., Jespersen, B., & Materna, P. (2010). *Procedural Semantics for Hyperintensional Logic. Foundations and Applications of Transparent Intensional Logic*. Berlin: Springer.

Materna, P. (1998). *Concepts and Objects*. Helsinki: Acta Philosophica Fennica.

Moschovakis, Y. N. (1994). Sense and Denotation as Algorithm and Value. In J. Väänänen & J. Oikkonen (Eds.), *Lecture Notes in Logic* (pp. 210–249). Berlin: Springer.

Raclavský, J. (2010). On Partiality and Tichý's Transparent Intensional Logic. *Hungarian Philosophical Review*, *54*, 120–128.

Salmon, N. (2010). Lambda in Sentences with Designators: An Ode to Complex Predication. *Journal of Philosophy*, *107*, 445–68.

Tichý, P. (1968). Smysl a procedura. *Filosofický časopis*, *16*, 222–232. (Translated as 'Sense and Procedure' in Tichý, 2004, pp. 77–92)

Tichý, P. (1969). Intensions in Terms of Turing Machines. *Studia Logica*, *26*, 7–25. (Reprinted in Tichý, 2004, pp. 93–109)

Tichý, P. (2004). *Collected Papers in Logic and Philosophy* (V. Svoboda, B. Jespersen, & C. Cheyne, Eds.). Prague: Filosofia, and Dunedin: University of Otago Press.

Marie Duží
VSB-Technical University Ostrava
The Czech Republic
E-mail: marie.duzi@gmail.com

Jakub Macek
VSB-Technical University Ostrava
The Czech Republic
E-mail: jakub.macek.0@gmail.com

Lukáš Vích
VSB-Technical University Ostrava
The Czech Republic
E-mail: luki.vich@gmail.com

Curry-Typed Semantics in Typed Predicate Logic

CHRIS FOX

Abstract: Various questions arise in semantic analysis concerning the nature of types. These questions include whether we need types in a semantic theory, and if so, whether some version of simple type theory (STT, Church, 1940) is adequate or whether a richer more flexible theory is required to capture our semantic intuitions. Propositions and propositional attitudes can be represented in an essentially untyped first-order language, provided a sufficiently rich language of terms is adopted. In the absence of rigid typing, care needs to be taken to avoid the paradoxes, for example by constraining what kinds of expressions are to be interpreted as propositions (Turner, 1992). But the notion of type is ontologically appealing. In some respects, STT seems overly restrictive for natural language semantics. For this reason it is appropriate to consider a system of types that is more flexible than STT, such as a Curry-style typing (Curry & Feys, 1958). Care then has to be taken to avoid the logical paradoxes. Here we show how such an account, based on the Property Theory with Curry Types (PTCT, Fox & Lappin, 2005), can be formalised within Typed Predicate Logic (TPL, Turner, 2009). This presentation provides a clear distinction between the classes of types that are being used to (i) avoid paradoxes (ii) allow predicative polymorphic types. TPL itself provides a means of expressing PTCT in a uniform language.

Keywords: Curry types, semantics, logic, intensionality

1 Introduction

In previous work the author has explored *Property Theory*, and issues that arise in intensionality, plurals, mass terms, discourse (Fox, 2000); *Property Theory with Curry Typing* (PTCT), and the representation of polymorphism, quantifier scoping, ellipsis (Fox & Lappin, 2005); "judgemental" proof-theoretic approaches to the semantics of imperatives (Fox, 2012a), obligations and permissions (Fox, 2012b), questions (Fox, 2013); as well as various related methodological issues (Fox, 2014; Fox & Turner, 2012).

The aim of this paper is to revisit earlier work on intensionality, but using a judgemental approach within a unifying framework. For natural language semantics we need fine-grained intensionality in order to capture the fact that we can have distinct beliefs about propositions with the same truth conditional, as well as having distinct questions with the same answerhood conditions, and distinct commands with the same satisfaction conditions. This requires something more fine-grained than Montague-style possible worlds semantics (Fox, 2000; Fox & Lappin, 2005). It is also appropriate to have flexible typing, to characterise the behaviour of polymorphic operators (and also to allow novel approaches to ellipsis and anaphora). This can be achieved by adopting a Curry-style approach to types (Curry & Feys, 1958). If possible, there should be some sensitivity to ontological questions, in particular we may wish to avoid an ontological collapse that results if everything is reduced to set-theoretic model-theory (Fox, 2014; Fox & Turner, 2012).

A central part of the approach adopted here is to regard logic as first-class citizen, and not merely a stepping-stone to a set-theoretic model. This is motivated by a desire to avoid ontological collapse, and treat ontological intuitions "faithfully". Avoiding ontological reduction helps to avoid unintended consequences.

1.1 An existing account

A pre-existing account for a logic with fine-grained intensionality and flexible typing is *Property Theory with Curry Typing* (Fox & Lappin, 2005). This consists of a "federation" of languages, in particular,

1. a language of wffs,
2. the untyped λ-calculus,
3. a language of types.

Wffs are used to express behaviour within a proof-theory. The λ-calculus provides the terms of the wffs. Intensional expressions are then *represented* by such terms.

Arguably, there are some æsthetic issues with the original presentation of PTCT. The multi-level syntax may appear rather clumsy. And the original tableaux-style presentation is hard to extend. There are also some other concerns, such as how a theory that focuses on propositions can be combined with analyses of non-propositional utterances and related phenomena, including imperatives, deontic expressions, questions and answers.

Curry-Typed Semantics in TPL

One way of resolving these issues is to adopt a cleaner, judgement-based proof-theoretic approach, were all the basic notions of theory are expressed in terms of a unified language of judgements.

1.2 Typed Predicate Logic

An alternative, unifying framework in which to formulate PTCT, and other theories, is provided by Typed Predicate Logic (TPL, Turner, 2008, 2009). TPL provides us with a minimal but expressive meta-theory. In some sense, it can be used in the same way that possible worlds is often used: it provides a generic framework, but without possible worlds problems (such as limited intensionality, and issues around ultimately reducing everything to sets). TPL allows us to express both syntactic and logical behaviour.

The basic judgements of TPL are as follows

1. T Type — T is a type

2. $t : T$ — t belongs to type T

3. p Prop — p is a proposition

4. p True (or just "p") — p is true

Such judgements appear with sequents of the form

$$\Gamma \vdash \Phi$$

where Φ is any judgement, and Γ a context. In TPL, both the syntactic and logical behaviour of a logic or language is expressed using sequent rules of the following form.

$$\frac{\Gamma_1 \vdash \Phi_1 \quad \Gamma_2 \vdash \Phi_2 \quad \ldots \quad \Gamma_n \vdash \Phi_n}{\Gamma \vdash \Phi}$$

Sometime we elide the contextual part of the judgement "$\Gamma_i \vdash$" when context Γ_i is the same for all judgements in a sequent rule.

We can exemplify the use of TPL by formulating some rules for propositions. Grammatical formation rules for conjunction can be expressed as follows,

$$\frac{\Gamma \vdash p \text{ Prop} \quad \Gamma \vdash q \text{ Prop}}{\Gamma \vdash p \wedge q \text{ Prop}} \wedge F$$

so if p and q are both propositions, then so is $p \wedge q$.

Logical rules for conjunction then characterise its classical behaviour.[1]

$$\frac{\Gamma \vdash p \text{ True} \quad \Gamma \vdash q \text{ True}}{\Gamma \vdash p \wedge q \text{ True}} \wedge + \qquad \frac{\Gamma \vdash p \wedge q \text{ True}}{\Gamma \vdash p \text{ True}} \wedge -$$

For universal quantification (typed), we can have the following formation rule

$$\frac{\Gamma, x : T \vdash p \text{ Prop}}{\Gamma \vdash \forall x \epsilon T \cdot p \text{ Prop}} \forall F$$

and the following logical rules.

$$\frac{\Gamma, x : T \vdash t \text{ True}}{\Gamma \vdash \forall x \epsilon T \cdot t \text{ True}} \forall + \qquad \frac{\Gamma \vdash \forall x \epsilon T \cdot t \text{ True} \quad \Gamma \vdash s : T}{\Gamma \vdash t[s/x] \text{ True}} \forall -$$

There are a number of ways in which we could show the consistency of a logic formulating in this way. For example, we might seek to show how the theory can be viewed as a conservative extension of another (sound) theory. Or we could build a model theory. In that case, the model theory is to be viewed as secondary to the logical theory; it is merely a tool to demonstrate properties of the logic, and not necessarily the intended framework for semantic analysis.

Here we shall see how PTCT can be reformulated within the framework of TPL. The proof-theoretic approach of TPL means that expressions can be thought of as being intensional "by default". Care will still need to be taken to avoid logical paradoxes.

2   PTCT in TPL

Theories are formulated in TPL by adding sequent rules. For PTCT we need a type system that allows for:

1. Curry-typed λ-calculus (for compositional semantics with flexible typing)

2. object-language representable types

3. type quantification (for polymorphic types, which avoids the need for "type-raising")

---

[1] Note that for a proper treatment, we would also need to give structural rules of *assumption*, *thinning* and *substitution* to allow the formation and logical rules to capture all aspects of the expected behaviour of the logical operators.

Curry-Typed Semantics in TPL

The nature of types, and object-level judgements, need to be constrained to avoid logical paradoxes, and to avoid impredicativity. To this end, PTCT has a simple stratification of types to avoid paradoxes and impredicativity while remaining weak.[2] The key kinds of types in PTCT can be classed as follows.

1. B — the basic type of individuals
2. Λ — "untyped" λ-calculus terms
3. τ — object level types
4. σ — quantifiable types

The type Λ is not an object level type. That is, it cannot appear *within* the language that we are formulated. The type τ is essentially a "type of types" but limited to those types that can appear within the expressions of the target language. As we shall see, the syntax will be further restricted so that there are no free-floating type judgements. The σ types are those over which we can quantify in the target language, as with polymorphic types. To avoid impredicativity, such types will exclude those that themselves involve quantification over types. The significance of some of these constraints will be considered again below (Section 3).

2.1 Structural rules

We first need to give housing keeping rules for "well-formedness" judgements. These are the usual rules for *assumption*, *thinning* and *substitution*, but specific to the case of formation rules in this TPL framework.[3]

$$\frac{\Gamma \vdash T \text{ Type}}{\Gamma, x : T \vdash x : T} \textit{Ass} \qquad \frac{\Gamma \vdash T \text{ Type} \quad \Gamma \vdash \Phi}{\Gamma, x : T \vdash \Phi} \textit{Thin}$$

$$\frac{\Gamma, x : T \vdash \Phi[x] \quad \Gamma \vdash t : T}{\Gamma \vdash \Phi[x/t]} \textit{Sub}$$

---

[2] We take it that weakness is a virtue particularly when it can be shown that relevant patterns of behaviour can be captured elegantly without resorting to more powerful formal machinery.

[3] We formulate these structural rules in a way where the context is explicitly narrowed down to just that part that is relevant to syntactic judgements, assuming that the syntax and logic are independent, but have chosen not to do so in order to simplify the exposition.

Here Φ stands for any type judgement. We also need versions of these structural rules for the logic behaviour, as follows.

$$\frac{\Gamma \vdash t \text{ Prop}}{\Gamma, t \vdash t} \text{ Ass}' \qquad \frac{\Gamma \vdash t \text{ Prop} \quad \Gamma \vdash \Phi}{\Gamma, t \vdash \Phi} \text{ Thin}'$$

$$\frac{t : \Lambda}{t =_\Lambda t} \text{ id}_1 \qquad \frac{t =_\Lambda s \quad \phi[t]}{\phi[s]} \text{ id}_2$$

Rules of this nature are not distinctive of PTCT, except that we have narrowed identity to just Λ terms.

2.2 λ-calculus

We can give formation/closure rules for λ-calculus/Λ, and identity rules to capture $\alpha\beta$-equivalence in TPL. These are of relevance to the structural rules for substitution. First we have the rules for the syntactic category of Λ.

$$\frac{s : \Lambda \quad t : \Lambda}{st : \Lambda} \Lambda_{\text{app}} \qquad \frac{\Gamma, x : \Lambda \vdash t : \Lambda}{\Gamma \vdash : \lambda x.t : \Lambda} \Lambda_{\text{abs}}$$

Function types for Λ/λ are governed by the following.

$$\frac{\Gamma, x : S \vdash t : T}{\Gamma \vdash \lambda x.t : S \Rightarrow T} \lambda_\Rightarrow + \qquad \frac{s : S \Rightarrow T \quad t : S}{st : T} \lambda_\Rightarrow -$$

And we have the usual α, β rules.[4]

$$\frac{s : \Lambda \quad t : \Lambda \quad r : \Lambda \quad t =_\Lambda r}{s =_\Lambda s[t/r]} \Lambda \alpha$$

$$\frac{\Gamma, x : \Lambda \vdash t : \Lambda \quad \Gamma \vdash s : \Lambda}{\Gamma \vdash (\lambda x.t)s =_\Lambda t[s/x]} \Lambda \beta$$

2.3 Propositions

TPL has a notion of proposition. But this is in the meta-theory. We need to give formation rules for the object level PTCT propositions (π) and the logical rules of truth for PTCT propositions.[5] Rather than give all the details of

---
[4]We could also add η.

[5]Note that the structural rules play a role in avoiding considering the truth of non-propositional expressions. There is insufficient space to demonstrate this, and it is left as an exercise for the interested reader.

Curry-Typed Semantics in TPL

all the PTCT connectives and quantifiers, we illustrate the nature of the rules using conjunction ($\wedge$) and universal quantification ($\forall$) as examples, together with the notions of intensional identity ($=$) and extensional equivalence ($\cong$). First we have the formation rules for expressions in π.

$$\frac{t : \pi \quad s : \pi}{(t \wedge s) : \pi} \wedge F$$

$$\frac{\Gamma, x : T \vdash t : \pi \quad \Gamma \vdash T : \tau}{\Gamma \vdash (\forall x \epsilon T \cdot t) : \pi} \forall F$$

$$\frac{t : T \quad s : T \quad T : \tau}{(t =_T s) : \pi} = F$$

$$\frac{t : T \quad s : T \quad T : \tau}{(t \cong_T s) : \pi} \cong F$$

Type T constrained to an object level type (i.e. T must be a type in τ).

The logical rules for conjunction and universal quantification follow.

$$\frac{s \quad t}{(s \wedge t)} \wedge + \quad \frac{(s \wedge t)}{s} \wedge -_1 \quad \frac{(s \wedge t)}{t} \wedge -_2$$

$$\frac{\Gamma, x : T \vdash t \quad \Gamma \vdash T : \tau}{\Gamma \vdash \forall x \epsilon T \cdot t} \forall + \quad \frac{\forall x \epsilon T \cdot t \quad s : T}{t[s/x]} \forall -$$

Intensional identity is a fine-grained notion. We can take it to be term identity in Λ.[6]

$$\frac{T : \tau \quad s : T \quad t : T \quad s =_\Lambda t}{s =_T t} =_T$$

Equivalence is a more coarse grained notion. For propositions it can be characterised as truth conditional equivalence.

$$\frac{s : \pi \quad t : \pi \quad s \leftrightarrow t}{s \cong_\pi t} \cong_\pi + \quad \frac{s \cong_\pi t}{s \leftrightarrow t} \cong_\pi -$$

The notion can be extended to "operational" extensional equivalence

$$\frac{x : S \quad f \cong_{S \Rightarrow T} g}{fx \cong_T gx} \cong_{S \Rightarrow T} -$$

As an example, properties will be equivalent if they apply to the same individuals.

---

[6]Intensional identity could be interpreted as operational identity of the λ-calculus, given that the latter can be taken to be a theory of computation.

2.4 Additional types

We can have rules for PTCT types, including formation and membership rules for polymorphic types and separation types.[7] Here are the rules for polymorphic types.

$$\frac{\Gamma, X : \sigma \vdash T : \tau}{\Gamma \vdash \Pi X \cdot T : \tau} \; \Pi F$$

A term t belongs to object-level type $\Pi X \cdot T$ iff t belongs to T for all "safe" values of X

$$\frac{\Gamma, X : \sigma \vdash t : T}{\Gamma \vdash t : \Pi X \cdot T} \; \Pi + \qquad \frac{t : \Pi X \cdot T \quad S : \sigma}{t : T[S/X]} \; \Pi -$$

Separation types are a form of subtype that combines types with propositions.

$$\frac{\Gamma \vdash T : \tau \quad \Gamma, x : T \vdash s : \pi}{\Gamma \vdash \{x \epsilon T \cdot s\} : \tau} \; SepF$$

A term t will be of the type $\{x \epsilon T \cdot s\}$ iff t is in T, and $s[x/t]$ is true.

$$\frac{\{x \epsilon T \cdot s\} : \tau \quad t : T \quad s[x/t]}{t : \{x \epsilon T \cdot s\}} \; Sep+$$

$$\frac{t : \{x \epsilon T \cdot s\}}{t : T} \; Sep-_1 \qquad \frac{t : \{x \epsilon T \cdot s\}}{s[x/t]} \; Sep-_2$$

2.5 An example

Without going into details of the compositional analysis, we can illustrate how a sentence may be represented within this framework. If we assume something like a Montagovian-like translation (Montague, 1973), *modulo* the use of PTCT, the sentence

> "*Every man loves a woman*"

can be given the representation

$$\forall x \epsilon B \cdot man'(x) \rightarrow \exists y \epsilon B \cdot woman'(y) \wedge loves'(x, y)$$

---

[7]Function types were introduced implicitly with the rules for λ-application and λ-abstraction.

Curry-Typed Semantics in TPL

where

$$man' : B \Rightarrow \pi$$
$$woman' : B \Rightarrow \pi$$
$$loves' : B \Rightarrow B \Rightarrow \pi$$

There is no need for "quoting" or an overt truth-operator, unlike original formulation of PTCT, and no need for Montague's "up" and "down" ($^\wedge, ^\vee$) operators (Montague, 1973, 1974).

In the case of the sentence

"John believes every man loves a woman"

we can have the representation

$$believes'(John')(\forall x \epsilon B \cdot man'(x) \rightarrow \exists y \epsilon B \cdot woman'(y) \wedge loves'(x,y))$$

where

$$believes' : B \Rightarrow (\pi \Rightarrow \pi)$$

Here, the "content" of the believe relation is *not* collapsed to a truth value, nor to a set of worlds, and again there is no need to "quote" the proposition, unlike in the original formulation of PTCT, nor use up and down operators to translate between intensional and extensional forms.

3 Paradoxes and impredicativity

In some respects, this formulation of PTCT in TPL can be viewed as a "flattened" version of a stratified logic, such as the one proposed by Turner (2005). Here we show how the simple layering of the types of PTCT, and other restrictions, help avoid paradoxes and impredicativity.

3.1 Paradoxes

We wish to avoid Russell-like paradoxes. One way is to block various forms of self-application, or arrange things so that their "truth" is never considered. Here we review two potentially problematic cases, (i) a universal type, and (ii) free-standing type-judgements.

In PTCT we have no universal type, and do not allow Λ to appear within PTCT expressions. To see why, consider expression rr where

$$r =_{\text{def}} \lambda x. \exists y \epsilon (\Lambda \Rightarrow P) \cdot x =_{\Lambda \Rightarrow P} y \wedge \neg xy$$

The problem is avoided by excluding Λ (or a universal type) from the object language.

PTCT also does not permit "Free-standing" type judgements. The reason can be seen by considering $r'r'$ where

$$r' =_{\text{def}} \{x\epsilon T \cdot \neg\, x\epsilon x\}$$

This problem is avoided if the use of $x\epsilon S$ is constrained. In both cases, other solutions to these potential pitfalls are possible.

PTCT is intended for the semantic interpretation of natural language. For this intended use, the lack of universal type and free-standing type-membership may not be an issue, as such expressions do not appear to be required to represent any natural utterance.

3.2 Impredicativity

In PTCT we have a form of explicit polymorphism ($\Pi X.T$). It is useful for conjunction, which we can type as $\Pi X.X \Rightarrow (X \Rightarrow X)$, and perhaps for and type general predicates (e.g. *"is fun"*: $\Pi X.X \longrightarrow \pi$, cf. Chierchia, 1982).

But polymorphic types can lead to an impredicative theory. If the type T is defined by the following,

$$T =_{\text{def}} \Pi X.S$$

we may wonder whether X range over T itself, the type that is being defined. If it does, then the definition is impredicative. In PTCT, this impredicativity is avoided by a two-level stratification:

1. τ — the types that can appear within PTCT

2. σ — the subclass of τ over which Π quantifies

The class of types σ excludes the polymorphic types themselves.

3.3 A truth predicate?

Finally, we can consider whether it is feasible to add a truth "predicate", or judgement, to the object language. Care would need to be taken to avoid logical paradoxes. If $\tau(t)$ were a PTCT expression corresponding to "t True", then we might wish to constrain it so that it can only form a proposition if t itself is a proposition (cf. Turner, 1990).

4 Other issues

Other issues we can reflect on are the consistency of PTCT, and the use of TPL as a more general unifying framework.

4.1 Interpretation and consistency

A modal for PTCT has already given elsewhere (Fox & Lappin, 2005). This treats intensional identity as inscriptional. An alternative is to think of intensions as procedures, so intensional identity is characterised as operational identity, and extensional equivalence is denotational equivalence. Note that PTCT already contains untyped λ-calculus, which can be interpreted as a model of computation. This proposal is explored in (Fox & Lappin, 2013).

4.2 Unified framework

One aim of formulating PTCT in TPL is to provide a unified framework that should make it easier to incorporate accounts of non-propositional phenomena. Previous work on imperatives (Fox, 2012a), obligations and permissions (Fox, 2012b), and questions (Fox, 2013) has appealed to logical characterisations involving a range of different judgements, including commanding, satisfying asking, answering. These might all be incorporated into TPL, or an extension of it.

Another unification that suggests itself is the formulation of natural language syntax in TPL, alongside its semantic interpretation in, for example, PTCT. The most natural grammar frameworks for might be along the lines of the Lambek calculus (Lambek, 1958), or some other type-logical grammar (Morrill, 1994; Steedman, 1993, 2000)—although there are no intrinsic constraints in TPL that would require us to adopt such formalisations. This would then provide an analysis of a fragment of natural language syntax and semantics in a uniform framework. The goal would be similar to that of Cooper (2012), where semantics and syntax are to be given a uniform analysis in terms of record types. But with TPL there would be no commitments or obligations to adhere to any particular kind of structured type.

5 Conclusions

We have reprised some core aspects of Property Theory with Curry Typing (PTCT), and presented a reformulation within Typed Predicate Logic (TPL). This gives an elegant presentation of the theory at a more abstract level, and avoids the need to invoke distinct levels of representation in the semantic analysis of natural language. PTCT is then captured within a single language in place of federation of languages for terms, types, and wffs. The notion of judgement in TPL may be adapted and extended to include judgements that are relevant for non-propositional utterances. The way in which the grammatical and logical rules are given in a uniform presentation in TPL may be extended to give rules governing both the syntax and semantic interpretation of natural language within a single unifying framework.

References

Chierchia, G. (1982). Nominalisation and Montague Grammar: A Semantics without Types for Natural Languages. *Linguistics and Philosophy, 5*, 303–354.

Church, A. (1940, June). A Formulation of the Simple Theory of Types. *The Journal of Symbolic Logic, 5*, 56–68.

Cooper, R. (2012). Type Theory and Semantics in Flux. In R. Kempson, T. Fernando, & N. Asher (Eds.), *Philosophy of Linguistics* (pp. 271–323). Amsterdam: Elsevier.

Curry, H. B., & Feys, R. (1958). *Combinatory Logic* (Vol. 1). Amsterdam: North Holland.

Fox, C. (2000). *The Ontology of Language*. Stanford: CSLI Publications.

Fox, C. (2012a). Imperatives: A Judgemental Analysis. *Studia Logica, 100*, 879–905.

Fox, C. (2012b). Obligations and Permissions. *Language and Linguistics Compass, 6*, 593–610.

Fox, C. (2013). Axiomatising Questions. In V. Punčochář & P. Švarný (Eds.), *Logica Yearbook 2012* (pp. 23–34). London: College Publications.

Fox, C. (2014). *The Meaning of Formal Semantics*. (To appear in P. Stalmaszczyk (Ed.), *Semantics and Beyond. Philosophical and Linguistic Investigations*, Frankfurt: Ontos Verlag.)

Fox, C., & Lappin, S. (2005). *Formal Foundations of Intensional Semantics*. Oxford: Blackwell.

Fox, C., & Lappin, S. (2013). *Type-theoretic Logic with an Operational Account of Intensionality*. (Manuscript)

Fox, C., & Turner, R. (2012). In Defense of Axiomatic Semantics. In Piotr Stalmaszczyk (Ed.), *Philosophical and Formal Approaches to Linguistic Analysis* (pp. 145–160). Frankfurt: Ontos Verlag.

Lambek, J. (1958). The Mathematics of Sentence Structure. *Mathematical Monthly*, 65, 154–169.
Montague, R. (1973). The Proper Treatment of Quantification in Ordinary English. In K. J. J. Hintikka, J. M. E. Moravcsik, & P. Suppes (Eds.), *Approaches to Natural Language* (pp. 221–242). Dordrecht: D. Reidel. (Also in Montague, 1974)
Montague, R. (1974). *Formal Philosophy* (R. H. Thomason, Ed.). New Haven: Yale University Press.
Morrill, G. (1994). *Type Logical Grammar: Categorial Logic of Signs*. Dordrecht: Kluwer Academic Publishers.
Steedman, M. (1993). Categorial Grammar (Tutorial Overview). *Lingua*, 90, 221–258.
Steedman, M. (2000). *The Syntactic Process*. Cambridge, MA: MIT Press.
Turner, R. (1990). *Truth and Modality for Knowledge Representation*. London: Pitman.
Turner, R. (1992). Properties, Propositions and Semantic Theory. In M. Rosner & R. Johnson (Eds.), *Computational Linguistics and Formal Semantics* (pp. 159–180). Cambridge: Cambridge University Press.
Turner, R. (2005). Semantics and Stratification. *Journal of Logic and Computation*, 15, 145–158.
Turner, R. (2008). Computable Models. *Journal of Logic and Computation*, 18, 283-318.
Turner, R. (2009). *Computable Models*. London: Springer.

Chris Fox
University of Essex
United Kingdom
E-mail: `foxcj@essex.ac.uk`

Relation Algebra throughout Galois Connections

RENATA DE FREITAS, LEANDRO SUGUITANI AND PETRUCIO VIANA[1]

Abstract: We present a set of non-equational axioms for relation algebra. Our axioms have three appealing characteristics: (1) they are based on the notion of *Galois connection*; (2) they display a remarkable parallelism between the Boolean (or *static*) and the non-Boolean (or *dynamic*) part of relation algebra; (3) they employ a mixture of equational reasoning, reasoning by equivalences, and reasoning by lattice theoretical inequalities, being very manageable in affording heuristics to derive the true equalities of relation algebra arithmetic.

Keywords: relation algebra, Galois connection, heuristics for proofs

1 Introduction

Relation algebra (RA) is an algebraic formalism, defined originally by Jónsson and Tarski (1948) (see also Chin & Tarski, 1951; Tarski, 1941), to play for binary relations the same role that Boolean algebra (BA) plays for unary relations (sets). Although it is arguable that RA has not fulfilled its role completely (Frias, 2002), it has been in use for investigating questions in many areas (Brink, Kahle, & Schmidt, 1997; Givant & Tarski, 1987; Schmidt & Ströhlein, 1993).

The definition of RA runs as follows. We say that an algebra

$$\mathfrak{A} = (A, +, \cdot, \mid, ^-, ^\smile, 0, 1, 1'),$$

where $+$, $\cdot$ and $\mid$ are binary operations, $^-$ and $^\smile$ are unary operations, 0, 1, and $1'$ are constants, is a *CT-relation algebra* if the axioms in Table 1 are satisfied, for all $r, s, t \in A$.

Axiom CT_1 is equivalent to a set of 3 equations. So, in the way it was originally defined, RA counts 8 operators and 11 axioms, being that (at least for a person versed in the basics of algebra and set theory) all the operators

---

[1] The work on this paper was supported by FAPESP, FAPERJ and CNPq.

(CT$_1$)	$(A, +, \cdot, ^-, 0, 1)$ is a Boolean algebra
(CT$_2$)	$(r \mid s) \mid t = r \mid (s \mid t)$
(CT$_3$)	$(r + s) \mid t = (r \mid t) + (s \mid t)$
(CT$_4$)	$r \mid 1' = 1' \mid r = r$
(CT$_5$)	$r^{\smile\smile} = r$
(CT$_6$)	$(r + s)^{\smile} = r^{\smile} + s^{\smile}$
(CT$_7$)	$(r \mid s)^{\smile} = s^{\smile} \mid r^{\smile}$
(CT$_8$)	$(r^{\smile} \mid (r \mid s)^-) + s^- = s^-$

Table 1: Chin-Tarski axioms for relation algebra.

have a very clear meaning when interpreted on relations and all but one of the axioms have a very clear algebraic meaning.

On the other hand, as outlined by Löwenheim (1915), and proved in detail by Givant and Tarski (1987), every mathematical problem—more specifically, every problem on the demonstrability of a mathematical statement from a given set of axioms—can be reduced to a problem of whether an equation can be proved from the CT-axioms for RA. This fact counts for the great expressive and proof powers of RA, but at the same time witnesses its high complexity, clarifying the intrinsic difficulty of determining whether a given equation holds in all relation algebras. As it was foreseen by Peirce (1883), and follows from the previous remarks, this problem is undecidable. So, there is no mechanical method to decide the consequences of the CT-axioms.

In this paper, we contribute to this problem, by presenting a set of axioms for RA that displays the following nice characteristics. First, all the axioms except one express the existence of a *Galois connection* between an adequate pair of partially ordered sets. Second, the axioms reveal a great deal of symmetry between the Boolean and the non-Boolean parts all relation algebras possess. Third, as it is emphasized in the papers (Aarts, 1992; Feijen, 2001; The Eindhoven Tuesday Afternoon Club, 1995), the existence of a Galois connection relating two operators may facilitate proving theorems, in the sense that the choice of a next step in a proof is guided by the directness inherited from the condition that the Galois connection imposes

Relation Algebra throughout Galois Connections 51

on the operators, a feature that we exemplify by presenting very detailed step-by-step proofs of arithmetical laws, from our axioms.

In Section 2, we review the basic facts on Galois connections that will be useful in what follows. In Section 3, we introduce a set of axioms for RA and prove that each but one of the axioms establishes a Galois connection between an adequately chosen pair of posets. We also apply the results from Section 2 to obtain some basic RA arithmetical laws. Finally, in Sections 4 and 5 we prove that the CT-axioms and the GC-axioms are equivalent.

2 Galois connections

In this section, we review the basics on Galois connections up to the point we need them. The reader interested in more on Galois connections may consult (Aarts, 1992; Erné, Koslowski, Melton, & Strecker, 2006), and the references therein.

Definition 1 *Let* $\mathfrak{P} = \langle P, \preceq_P \rangle$ *and* $\mathfrak{Q} = \langle Q, \preceq_Q \rangle$ *be posets endowed with functions* $f\colon P \to Q$ *and* $g\colon Q \to P$. *We say that* $\langle \mathfrak{P}, \mathfrak{Q}, f, g \rangle$ *is a* Galois connection *if*

$$fx \preceq_Q y \text{ iff } x \preceq_P gy, \tag{1}$$

for all $x \in P$ *and* $y \in Q$. *In this case, we say that* f *is the* lower *adjoint of* g, *and that* g *is the* upper *adjoint of* f.

From (1), we can warrant useful properties of f and g—related to partial orderings $\preceq_P$ and $\preceq_Q$—both when applied individually, as well as when their applications are nested. These are summarized in the next proposition:

Proposition 1 *Let* $\langle \mathfrak{P}, \mathfrak{Q}, f, g \rangle$ *be a Galois connection. Then, for every* $\{x, y\} \cup \{x_i : i \in I\} \subseteq P$, *and* $\{u, v\} \cup \{u_i : i \in I\} \subseteq Q$, *we have:*
(a) $x \leq_P gfx$.
(b) $fgu \leq_Q u$.
(c) *If* $x \leq_P y$, *then* $fx \leq_Q fy$.
(d) *If* $u \leq_Q v$, *then* $gu \leq_P gv$.
(e) *If* $\bigvee \{x_i : i \in I\}$ *exists, then* $f \bigvee \{x_i : i \in I\} =_Q \bigvee \{fx_i : i \in I\}$.
(f) *If* $\bigwedge \{u_i : i \in I\}$ *exists, then* $g \bigwedge \{u_i : i \in I\} =_P \bigwedge \{gy_i : i \in I\}$.
(g) *If* $0_P \leq_P x$, *for every* $x \in P$, *then* $f0_P \leq_Q y$, *for every* $y \in Q$.
(h) *If* $y \leq_Q 1_Q$, *for every* $y \in Q$, *then* $x \leq_P g1_Q$, *for every* $x \in P$.

Items (e) and (f) of Proposition 1 are necessary conditions for a function between posets to have upper and lower adjoints. They might be stretched as follows:

Proposition 2 *Let $\mathfrak{P} = \langle P, \preceq_P \rangle$ and $\mathfrak{Q} = \langle Q, \preceq_Q \rangle$ be posets.*
(a) *Let $f\colon P \to Q$ be a function such that, for every $y \in Q$, the set $\{z \in P : fz \leq_Q y\}$ has a lub in P. Then, f has an upper adjoint iff $f \bigvee \{z \in P : fz \preceq_Q y\} \leq_Q y$ and f preserves $\preceq_P$.*
(b) *Let $g\colon Q \to P$ be a function such that, for every $x \in P$, the set $\{z \in Q : x \leq_P gz\}$ has a glb in Q. Then, g has a lower adjoint iff $x \preceq_P g \bigwedge \{z \in P : fz \preceq_Q y\}$ and g preserves $\preceq_Q$.*

3 Relation algebra axioms throughout Galois connections

In this section, we introduce another set of axioms for RA. After that, we prove that most of the axioms establish Galois connections between adequately chosen posets. We finish the section by presenting some basic arithmetical consequences of the axioms, based on Proposition 1.

We say that an algebra provided with a binary relation

$$\mathfrak{A} = (A, \cdot, \mid, ^-, ^\perp, 0, 0', \leq),$$

where $\cdot$ and $\mid$ are binary operations, $^-$ and $^\perp$ are unary operations, 0 and $0'$ are constants and $\leq$ is a partial order, is a *GC-relation algebra* if the axioms in Table 2 are satisfied, for all $r, s, t \in A$.

This axiomatization is based on six primitive notions, which can be divided in two groups. The first group is composed by the notions $\cdot$ (parallelization), $^-$ (complementation) and 0 (abort) which form the static part of the algebra. Whereas to the second group belong the notions $\mid$ (sequentialisation), $^\perp$ (estrangement), and $0'$ (differentiation) which form the dynamical part of the algebra. The B in the static axioms is for Boole, the P in the dynamic axioms is for Peirce, the S is for Schröder, and the D for Dijkstra.

Axioms (B_3) and (B_4) are related in the sense that both deal with occurrences of the operator $^-$. The former displays an equivalence for the cases in which $^-$ occurs as the main operator on the left hand side of an inclusion, and the later displays a similar equivalence for the cases in which $^-$ occurs as the main operator on the right hand side. Entirely analogous observations hold for the pair (B_5) and (S), dealing with occurrences of $\cdot$, and the pair (P_3) and (P_4), dealing with occurrences of $^\perp$. Axioms (P_5) and (D) deal with occurrences of $\mid$. Moreover, axioms (B_i) and (P_i), $1 \leq i \leq 5$, have a close relationship between their forms, they also come in pairs. In fact, they are related in the sense that each axiom (P_i) can be obtained from the

Static axioms

(B$_1$) $r \cdot (s \cdot t) \leq (r \cdot s) \cdot t$ and $(r \cdot s) \cdot t \leq r \cdot (s \cdot t)$
(B$_2$) $0^- \cdot r \leq r$ and $r \leq r \cdot 0^-$
         $r \cdot 0^- \leq r$ and $r \leq 0^- \cdot r$
(B$_3$) $r^- \leq s$ iff $s^- \leq r$
(B$_4$) $r \leq s^-$ iff $s \leq r^-$
(B$_5$) $r \cdot s \leq t$ iff $s \leq (t^- \cdot r)^-$
(S) $r \leq s \cdot t$ iff $r \leq s$ and $r \leq t$

Dynamic axioms

(P$_1$) $r \mid (s \mid t) \leq (r \mid s) \mid t$ and $(r \mid s) \mid t \leq r \mid (s \mid t)$
(P$_2$) $0'^{\perp} \mid r \leq r$ and $r \leq r \mid 0'^{\perp}$
         $r \mid 0'^{\perp} \leq r$ and $r \leq 0'^{\perp} \mid r$
(P$_3$) $r^{\perp} \leq s$ iff $s^{\perp} \leq r$
(P$_4$) $r \leq s^{\perp}$ iff $s \leq r^{\perp}$
(P$_5$) $r \mid s \leq t$ iff $s \leq (t^{\perp} \mid r)^{\perp}$
(D) $(r^{\perp} \mid s^{\perp})^{\perp} \leq (s^- \mid r^-)^-$ and $(s^- \mid r^-)^- \leq (r^{\perp} \mid s^{\perp})^{\perp}$

Table 2: Galois connection axioms for relation algebra.

corresponding axiom (B$_i$), and vice-versa, by interchanging the occurrences of $\cdot$, $^-$, and 0 by occurrences of $\mid$, $^{\perp}$, and 0', respectively.

As usual, we interpret the conjunction of two reversibly related inequalities, $r \leq s \wedge s \leq r$, as an equality, $r = s$. In this way, Axiom (B$_1$) warrants that $\cdot$ is associative, Axiom (B$_2$) warrants that 0^- is a neutral element for $\cdot$, and Axioms (B$_3$) and (B$_4$) warrants that $^-$ is an involution. Hence, axioms (B$_1$),...,(B$_4$) warrant that, for any GC-relation algebra $\mathfrak{A}$, the associated algebraic system $\langle A, \cdot, 0^- \rangle$ is an involuted monoid. From the symmetry emphasized above, we also conclude that axioms (P$_1$),...,(P$_4$) warrant that the associated algebraic system $\langle A, \mid, 0'^{\perp} \rangle$ also is an involuted monoid.

Now, we prove that each one of the axioms (B$_3$), (B$_4$), (B$_5$), (S), (P$_3$),

(P$_4$), and (P$_5$) establishes a Galois connection between posets. In what follows, $\mathfrak{A}$ is an arbitrary GT-relation algebra.

Proposition 3 *Axioms* (B$_3$), (B$_4$), (B$_5$) *and* (S) *establish Galois connections.*

Proof. For (B$_3$), let $P = Q = A$, $\preceq_P$ is $\geq$, and $\preceq_q$ is $\leq$. Besides, define functions $f\colon P \to Q$ and $g\colon Q \to P$ by setting $fr = gr = r^-$, for every $r \in A$.

The proof for (B$_4$) is analogous, by interchanging the roles of $\mathfrak{P}$ and $\mathfrak{Q}$ above.

For (B$_5$), given $r \in A$, let $P = Q = A$, $\preceq_P$ is $\leq$, and $\preceq_q$ is $\leq$. Besides, define functions $f\colon P \to Q$ and $g\colon Q \to P$ by setting $fs = r \cdot s$ and $gt = (t^- \cdot r)^-$, for all $s, t \in A$. Hence, for all $s, t \in A$, we have $fs \preceq_Q t$ iff $r \cdot s \leq t$ iff, by (B$_5$), $s \leq (t^- \cdot r)^-$ iff $s \preceq_P gs$.

For (S), let $P = A$, $Q = A \times A$, $\preceq_P$ is $\leq$, and $\preceq_q$ is the order induced by $\leq$ on $A \times A$. Besides, define functions $f\colon P \to Q$ and $g\colon Q \to P$ by setting $fr = (r, r)$ and $g(s, t) = s \cdot t$, for all $r, s, t \in A$. Hence, for all $r, s, t \in A$, we have $fr \leq (s, t)$ iff $(r, r) \leq (s, t)$ iff $r \leq s$ and $r \leq t$ iff, by (S), $r \leq s \cdot t$ iff $r \preceq_P g(s, t)$. □

Since (P$_3$), (P$_4$), and (P$_5$) can be obtained from (B$_3$), (B$_4$), and (B$_5$) by interchanging the occurrences of $\cdot$, $^-$, and 0 by occurrences of $|$, $^\perp$, and $0'$, respectively, Proposition 4 follows without proof.

Proposition 4 *Axioms* (P$_3$), (P$_4$), *and* (P$_5$) *establish Galois connections.*

The parallelism between the static and the dynamic parts of the GC-axioms is broken by axioms (S) and (D). As we already mentioned, axioms (B$_i$) and (P$_i$), $1 \leq i \leq 5$, come in pairs, but we do not have a dynamic axiom analogous to (S). Proposition 2 can be used to explain why this is so.

Let $\mathfrak{A}$ be the full set relation algebra on $\{a, b\}$, $\preceq$ be the order on $A \times A$ induced by $\subseteq$, and $g\colon A \times A \to A$ be a function defined by $g(s, t) = s \circ t$. Then $\bigcap\{(s, t) \in A \times A : r \subseteq s \circ t\}$ exists for every $r \in A$. Now, given $r = \{(a, b)\}$, we have that $\bigcap\{(s, t) \in A \times A : r \subseteq s \circ t\} = \{(\emptyset, \emptyset)\}$, since, for instance, $r \subseteq \{(a, a)\} \circ \{(a, b)\}$ and $r \subseteq \{(a, b)\} \circ \{(b, b)\}$. Then $r \not\subseteq g\bigcap\{(s, t) \in A \times A : r \subseteq s \circ t\}$ and, by Proposition 2, the function $g\colon A \times A \to A$ such that $g(s, t) = s \circ t$ does not have a lower adjoint. Hence, we have that there is no dynamic axiom analogous to (S), because such an axiom, just like (S), would establish a Galois connection and, then, g would have a lower adjoint.

By putting Propositions 1, 3, and 4 together, we have the following useful arithmetical properties.

Corollary 1 *For every r, s, t in a GC-relation algebra, we have:*
(a) $r^{--} \leq r \wedge r \leq r^{--}$.
(b) $r \leq s \Longrightarrow s^- \leq r^-$.
(c) $r \cdot (s^- \cdot r)^- \leq s$.
(d) $s \leq t \Longrightarrow r \cdot s \leq r \cdot t$.
(e) $r \cdot s \leq s$.
(f) $r \cdot s \leq r$.
(g) $r \leq s \wedge t \leq u \Longrightarrow r \cdot t \leq s \cdot u$.
(h) $r^{\perp\perp} \leq r \wedge r \leq r^{\perp\perp}$.
(i) $r \leq s \Longrightarrow s^\perp \leq r^\perp$.
(j) $s \leq t \Longrightarrow r \mid s \leq r \mid t$.

Corollary 1(b) warrants that $^-$ is antitonic. Since its proof is based on the Galois connection given by (B_3) and (B_4), we also have the analogous Corollary 1(i), warranting that $\perp$ is antitonic as well. Corollary 1(d) warrants the left-monotonicity of $\cdot$. Since its proof is based on the Galois connection given by (B_5), we also have the analogous Corollary 1(j), warranting the left-monotonicity of $\mid$. Corollary 1(g) warrants that $\cdot$ is monotonic. But since its proof is based on the Galois connection given by (S), we do not have an immediate analogous warranting that $\mid$ is monotonic. However, an interplay of axioms (P_4) and (P_5) will give us the right-monotonicity of $\mid$.

Observation We would like to emphasize that in what follows we adopt the *calculational proof style* introduced by W. H. J. Feijen (cf. Aarts, 1992; Feijen, 2001). In doing this, we are taking advantage of the shape of the GC-axioms. Usually each step of the proof is guided by an exam of the previous step in a tentative of matching it to some instance of some axiom. Although this procedure is far from deterministic, in much cases it leaves us with hardly any manipulative freedom.

Proposition 5 *For all r, s, t in a GC-relation algebra, we have $r \leq s \Longrightarrow r \mid t \leq s \mid t$.*

Proof. Given arbitrary $r, s, t \in A$, we have:

$r \mid t \leq s \mid t$
$\Uparrow$ by Corollary 1(h)
$r \mid t \leq (s \mid t)^{\perp\perp}$
$\Uparrow$ by (P$_4$)
$(s \mid t)^{\perp} \leq (r \mid t)^{\perp}$
$\Uparrow$ by Corollary 1(h)
$(s \mid t)^{\perp} \leq (r^{\perp\perp} \mid t)^{\perp}$
$\Uparrow$ by (P$_5$)
$t \mid (s \mid t)^{\perp} \leq r^{\perp}$
$\Uparrow$ by (P$_4$)
$r \leq (t \mid (s \mid t)^{\perp})^{\perp}$
$\Uparrow$ by the hypothesis ($r \leq s$)
$s \leq (t \mid (s \mid t)^{\perp})^{\perp}$
$\Uparrow$ by Corollary 1(h)
$s \leq (t^{\perp\perp} \mid (s \mid t)^{\perp})^{\perp}$
$\Uparrow$ by (P$_5$)
$(s \mid t)^{\perp} \mid s \leq t^{\perp}$
$\Uparrow$ by (P$_4$)
$t \leq ((s \mid t)^{\perp} \mid s)^{\perp}$
$\Uparrow$ by (P$_5$)
$s \mid t \leq s \mid t$

This concludes the proof. □

4 The Boolean reduct

In this section, we prove that axioms (B$_3$), (B$_4$), (B$_5$) and (S) force the monoid $\langle A, \cdot, 0^- \rangle$ to be a Boolean lattice (i.e. a complemented distributive bounded lattice).

First, we define the missing operators by taking

$$r + s ::= (r^- \cdot s^-)^- \quad \text{and} \quad 1 ::= 0^-,$$

for all $r, s \in A$.

Proposition 6 *Let* $\mathfrak{A} = (A, \cdot, \mid, ^-, ^\perp, 0, 0', \leq)$ *be a GC-relation algebra. Then axioms* (B$_3$), (B$_4$), (B$_5$) *and* (S) *warrant that*

$$BA\mathfrak{A} = (A, +, \cdot, ^-, 0, 1, \leq)$$

is a complemented distributive bounded lattice.

Due to lack of space, we omit the proof.

5 Equivalence between the CT and the GC axioms

In this section, we prove that the CT-axioms (cf. Table 1) and the GC-axioms (cf. Table 2) are equivalent. First, we define the CT primitives which are not part of the GC repertoire. After that, we prove that the CT-axioms can be derived from the GC-axioms. Then we do the other way around, defining the GC primitives which are not part of the CT repertoire and proving that GC-axioms can be derived from the CT-axioms.

To prove axioms (CT$_2$), ..., (CT$_8$) we use some further GC-theorems, which show a deeper relationship between the static and dynamics parts of the GC-relation algebras.

Proposition 7 *For all r, s in a GC-relation algebra, we have:*
(a) $0'^\perp = 0'^-$.
(b) $r \mid s \leq 0'$ *iff* $r \leq s^\perp$.
(c) $\forall x (r \mid x^\perp \leq 0'$ *iff* $s \mid x^\perp \leq 0') \implies r = s$.

Due to lack of space, we omit the proofs.

Item (a) is an expected relationship between $^-$, $^\perp$ and $0'$. Items (b) and (c) are inspired by the implication (2) $\forall x (r \leq x$ iff $s \leq x) \implies r = s$, which holds on all partially ordered sets. But, since we are in a Boolean algebra, (2) can be rewritten as $\forall x (r \cdot x^- \leq 0$ iff $s \cdot x^- \leq 0) \implies r = s$. which is analogous to (c), due to the close parallelism between the static and the dynamic parts of the GC-relation algebra.

Finally, we are ready to prove the important equivalence showing that complement and estrangement commute. In what follows, we refer to aplications of Proposition 6 simply as BA.

Proposition 8 *For every r in a GC-relation algebra, we have $r^{\perp -} = r^{-\perp}$.*

Proof. By Proposition 7, it suffices to prove that $r^{-\perp} \mid x \leq 0'$ if, and only if $r^{\perp -} \mid x \leq 0'$, for an arbitrary x. To this, we have:

$r^{-\perp} \mid x \leq 0'$
$\Updownarrow$ by (P$_4$), Corollary 1(i), and Proposition 5
$(r^{-\perp} \mid x^{\perp\perp})^{\perp\perp} \leq 0'$
$\Updownarrow$ by (D)
$(x^{\perp -} \mid r^{--})^{-\perp} \leq 0'$
$\Updownarrow$ by (P$_3$)
$0'^{\perp} \leq (x^{\perp -} \mid r^{--})^{-}$
$\Updownarrow$ by (B$_3$), and Proposition 5
$x^{\perp -} \mid r \leq 0'$
$\Updownarrow$ by Proposition 7
$x^{\perp -} \leq r^{\perp}$
$\Updownarrow$ by (B$_3$) and BA
$r^{\perp -} \leq x^{\perp}$
$\Updownarrow$ by Proposition 7
$r^{\perp -} \mid x \leq 0'$.

This completes the proof. $\square$

We already defined the missing Booleans and proved axiom (CT$_1$) in Section 4. To the other missing operators, we set:

$$r^{\vee} ::= r^{\perp -} \quad \text{and} \quad 1' ::= 0'^{\perp}.$$

Proposition 9 *Let $\mathfrak{A} = (A, \cdot, \mid, ^{-}, ^{\perp}, 0, 0')$ be a GC-relation algebra. Then $\mathrm{CT}\mathfrak{A} = (A, +, \cdot, \mid, ^{-}, ^{\vee}, 0, 1, 1')$ is a CT-relation algebra.*

Proof. We proved axiom (CT$_1$) in Section 4. We treat the other axioms as follows. Axiom (CT$_2$) is (P$_1$). Axiom (CT$_3$) follows from Proposition 4. Axiom (CT$_4$) is (P$_2$), according to the definition $0'^{\perp} = 1'$. To prove (CT$_5$), we have:

$r^{\smile\smile} = r$
$\Updownarrow$ by the definition of $\smile$
$r^{\perp-\perp-} = r$
$\Updownarrow$ by (P$_3$) and (P$_4$)
$(r^\perp)^- = (r^-)^\perp$

This last equality is true by Proposition 8. To prove (CT$_6$) we prove the stronger result that, for all r, s in a GC-relation algebra, $r^{\smile} \leq s$ iff $r \leq s^{\smile}$.

This is the same as proving that $\smile$ establishes a Galois connection. So, (CT$_6$) follows by a direct application of Proposition 2. To this, we have:

$r^{\smile} \leq s$
$\Updownarrow$ by the definition of $\smile$
$r^{\perp-} \leq s$
$\Updownarrow$ by (B$_3$)
$s^- \leq r^\perp$
$\Updownarrow$ by (P$_4$)
$r \leq s^{-\perp}$
$\Updownarrow$ by Proposition 8
$r \leq s^{\perp-}$
$\Updownarrow$ by the definition of $\smile$
$r \leq s^{\smile}$

To prove (CT$_7$), we have:

$(r \mid s)^{\smile} = s^{\smile} \mid r^{\smile}$
$\Uparrow$ by the definition of $\smile$
$(r \mid s)^{\perp-} = s^{\perp-} \mid r^{\perp-}$
$\Uparrow$ by BA
$(r \mid s)^\perp = (s^{\perp-} \mid r^{\perp-})^-$
$\Uparrow$ by (D)
$(r \mid s)^\perp = (r^{\perp\perp} \mid s^{\perp\perp})^\perp$
$\Uparrow$ by Corollary 1
$(r \mid s)^\perp = (r \mid s)^\perp$.

To prove (CT$_8$) we have:

$(r^\vee \mid (r \mid s)^-) + s^- = s^-$
⇑ by the definition of $+$
$((r^\vee \mid (r \mid s)^-)^- \cdot s^{--})^- = s^-$
⇑ by BA
$(r^\vee \mid (r \mid s)^-)^- \cdot s = s$.

By BA, we have $(r^\vee \mid (r \mid s)^-)^- \cdot s \leq s$. The other inequality can be proved as follows.

$s \leq (r^\vee \mid (r \mid s)^-)^- \cdot s$
⇑ by (S)
$s \leq (r^\vee \mid (r \mid s)^-)^-$
⇑ by (B$_3$)
$r^\vee \mid (r \mid s)^- \leq s^-$
⇑ by (P$_5$)
$(r \mid s)^- \leq (s^{-\perp} \mid r^\vee)^\perp$
⇑ by the definition of $^\vee$ and Proposition 8
$(r \mid s)^- \leq (s^{-\perp} \mid r^{-\perp})^\perp$
⇑ by (D)
$(r \mid s)^- \leq (r^{--} \mid s^{--})^-$,

which also is true by BA. □

To prove that the CT-axioms imply the GC-axioms, we define the missing operators as:

$$r^\perp = r^{-\vee} \quad \text{and} \quad 0' = 1'^-.$$

Proposition 10 *Let* $\mathfrak{A} = (A, +, \cdot, \mid, ^-, ^\vee, 0, 1, 1')$ *be a CT-relation algebra. Then* $\text{GC}\mathfrak{A} = (A, \cdot, \mid, ^-, ^\perp, 0, 0')$ *is a GC-relation algebra.*

Proof. Axioms (B$_1$), ..., (S) are very well known consequences of (CT$_1$). Axiom (P$_1$) is (CT$_2$). Axiom (P$_2$) follows from (CT$_1$) and from (Chin & Tarski, 1951, Corollary 1.4, Corollary 1.5). Axioms (P$_3$) and (P$_4$) follow from (CT$_1$), (CT$_5$), and from (Chin & Tarski, 1951, Theorem 1.8). Axiom (P$_5$) follows from (CT$_1$), (CT$_5$), (CT$_7$), and from (Chin & Tarski, 1951,

Theorem 1.10, Theorem 2.1). Finally, (D) follows from (CT_5), (CT_7), and from (Chin & Tarski, 1951, Theorem 1.10). □

References

Aarts, C. J. (1992). *Galois Connections Presented Calculationally*. PhD Thesis, Eindhoven University of Technology.

Brink, C., Kahle, W., & Schmidt, G. (Eds.). (1997). *Relational Methods in Computer Science*. New York: Springer.

Chin, L. H., & Tarski, A. (1951). Distributive and Modular Laws in the Arithmetics of Relation Algebras. *University of California Publications in Mathematics*, *1*, 341–384.

Erné, M., Koslowski, J., Melton, A., & Strecker, G. E. (2006). A Primer on Galois Connections. *Annals of the New York Acadademy of Science*, *704*, 103–125.

Feijen, W. (2001). The Joy of Formula Manipulation. *Information Processing Letters*, *77*, 89–96.

Frias, F. M. (2002). *Fork Algebras in Algebra, Logic and Computer Science*. Singapore: World Scientific.

Givant, S., & Tarski, A. (1987). *A Formalization of Set Theory without Variables*. Providence: American Mathematical Society.

Jónsson, B., & Tarski, A. (1948). Representation Problems for Relation Algebras. *Bulletin of the American Mathematical Society*, *54*, 80.

Löwenheim, L. (1915). Über Möglichkeiten im Relativkalkül. *Mathematische Annalen*, *76*, 447–470.

Peirce, C. S. (1883). The Logic of Relatives (Note B). In *Studies in Logic, by Members of the Johns Hopkins University*. Boston: Little Brown and Company.

Schmidt, G., & Ströhlein, T. (1993). *Relations and Graphs*. Berlin, Heidelberg: Springer.

Tarski, A. (1941). On the Calculus of Relations. *Journal of Symbolic Logic*, *6*, 73–89.

The Eindhoven Tuesday Afternoon Club. (1995). Constructing the Galois Adjoint. *Information Processing Letters*, *53*, 137–139.

Renata de Freitas
Fluminense Federal University
Brazil
E-mail: renatafreitas@id.uff.br

Leandro Suguitani
University of Campinas
Brazil
E-mail: leandrosuguitani@gmail.com

Petrucio Viana
Fluminense Federal University
Brazil
E-mail: petrucio@cos.ufrj.br

Modeling Truly Dynamic Epistemic Scenarios in a Partial Version of DEL

JENS ULRIK HANSEN[1]

Abstract: Dynamic Epistemic Logic is claimed to be a dynamic version of epistemic logic. While this being true, there are several dynamical aspects that cannot be reasoned about in Dynamic Epistemic Logic. When a scenario is fixed and a possible world model representing the scenario is constructed, the possible future ways the system can evolve are in some sense already determined. For instance no new agents can enter the scenario and no new propositional facts can become relevant. This modeling perspective is the main motivation for the partial version of Dynamic Epistemic Logic introduced in this paper, which in particular, allows for the set of agents and the set of propositional variables to change.

Keywords: dynamic epistemic logic, partial logic, epistemic modeling, logical modeling

1 Introduction

Broadly construed, the field of logic can be viewed in several different ways, for instance, as a normative science of reasoning. However, the view presupposed in this paper is that logic is simply a formal modeling tool on par with other mathematical frameworks such as Partial Differential Equations. When modeling a scenario using logic there are two different approaches one could take: Either one could start with describing the scenario in a formal language; or one could start by setting up a formal semantic model. Derived information about the scenario in question can then be obtained by either theorem proving or model checking. These two distinct approaches to modeling both have their advantages and disadvantages and thus, bridging them seems like a worthwhile enterprise. Making the "language approach" more like the "modeling approach" is somewhat achieved by Hybrid Logic

---

[1] The work reported in this paper was supported by an individual postdoc grant from the Carlsberg Foundation entitled "Enhancing fitness and flexibility of logic-based models of knowledge and beliefs".

(Areces & ten Cate, 2007; Blackburn, 2000), whereas making the modeling approach more like the language approach can be realized by moving to partial logic.

Starting from the model approach, when modeling epistemic scenarios (using Epistemic Logic) we are faced with the task of coming up with suitable possible world models of the scenarios.[2] As argued by van Benthem (2009) there is not much systematic science to this – we are left with "the art of modeling". However, if we are concerned with dynamic or evolving scenarios the art of modeling has two aspects: (i) Coming up with the initial model, and (ii) Specifying how the initial model evolves. The claim to success of Dynamic Epistemic Logic (DEL) (van Ditmarsch, van der Hoek, & Kooi, 2008) is exactly that it makes (ii) into a systematic science for completely described scenarios. However, (i) remains as an art. Thus, developing a partial version of DEL can serve two purposes: Making coping with partially described scenarios possible, and making (i) more into a science than an art.

Even though DEL is claimed to be a dynamic version of Epistemic Logic there are several dynamical aspects that cannot be reasoned about in DEL. In DEL, when a possible world model is constructed to represent an epistemic scenario the possible future ways the system can evolve are in some sense already determined. For instance, no new agents can enter the scenario and no new propositional facts can become relevant. The set of agents and the set of propositional variables are fixed at the start of the modeling and remain constant throughout any future developments of the system. In this sense, there is an inherent "closed-system assumption". However, in a truly dynamic world new agents may come in having relevant knowledge, or the agents may become aware of relevant issues that were previously considered irrelevant. Attempts to model such dynamics have already been made, for instance in awareness logic (Fagin & Halpern, 1988; van Ditmarsch & French, 2009, 2011). However, these attempts assume that all the new agents and all the propositional facts that the agents might become aware of in the future are specified initially. Thus, as a modeler, one still has to account for all future developments of the scenario at the outset. In this paper, these insufficiencies are remedied by developing a partial version of DEL.

---

[2]For simplicity, other semantics for Epistemic Logic, than possible world semantics, is ignored in this paper.

Let us consider an example of an epistemic scenario of the kind typical modeled in DEL: *Three friends, Arnold, Bruce, and Chuck are having a friendly game of Texas Hold'em poker at Bruce's place. Arnold shuffles the cards and deals them.* The friends' knowledge in the game can be represented by a possible world model, where each world corresponds to a particular deal of the cards (each player receives two cards) and a player cannot distinguish between two deals of cards if he holds the same two cards in each of them. We introduce propositional variables for each player holding a particular card and can then say things like "Chuck holds the two of hearts" or "Bruce does not know that Chuck holds the two of hearts". Continuing the story: *The betting begins with no one folding, after which Chuck, being the dealer, deals the flop (puts three cards face-up on the table).* This event reduces the set of possible world by excluding all the deals where one of the friends have one of the cards on the table, and can easily be modeled using a public announcement modality (the simplest form of dynamics in DEL). An event, such as *Arnold shows one of his cards to Bruce*, can also be modeled in DEL using a more complex type of modality based on so called event models. In the resulting model several indistinguishability links for Bruce will be removed since now he has no uncertainty about Arnold's cards.

Now assume that *Bruce's wife walks in*. How to update the model? That of course depends on whether we think it has any influence on the game, but one could easily imagine that Bruce and his wife have a secret way of communicating and that she is able to see Chuck's cards. Now suddenly, there is a new agent present that has information about Chuck's cards and she may or may not have communicated them to Bruce. Finally, imagine that *Chuck announces that he has the ability to see through cards, but he never uses it unless he has an incredibly bad hand.* This introduces a new atomic proposition "Chuck has looked at Arnold's and Bruce's cards" to which Arnold and Bruce are uncertain, but Arnold and Bruce knows that this proposition is only true if Chuck has an incredibly bad hand. This new proposition becomes relevant if Arnold, for instance, has a chance to learn Chuck's cards (for instance by looking at them while Chuck is out for more beers).

What this (un)realistic example hopefully shows is that even though we could have included the extra atomic propositions and the extra agents from the beginning, it leaves the modeler with an extremely difficult job of planning for all possible future unlikely, but still possible, events. Thus, a partial epistemic logic with guidelines for how to transform partial possible world models under events of the above type (new agents and atomic facts entering the scenario) is certainly valuable from a modeling perspective.

The structure of the paper is as follows: In Section 2 we recapitulate a partial version of Epistemic Logic. Then, we move on to introduce a novel partial version of DEL in Section 3. In Section 4 we discuss *reduction laws* for the newly introduced partial version of DEL, which are the counterparts of reduction axioms in classical DEL. Finally, a conclusion and future research are given in Section 5.

2 Partial Epistemic Logic and possible world models

In this section, we introduce a partial version of Epistemic Logic, which is already well-known in the literature (Jaspars, 1994; Jaspars & Thijsse, 1996; Thijsse, 1992). The main idea is to define partial possible world models as models where each world is equipped with only a partial valuation of propositional variables.

On the syntactic side, Partial Epistemic Logic resembles ordinary Epistemic Logic completely. In other words, we assume a set of agents $\mathbb{A}$ and a countable infinite set of propositional variables PROP. Then, the syntax of the language is given in the following way:

$$\varphi ::= p \mid \top \mid \neg\varphi \mid \varphi \wedge \varphi \mid K_a\varphi ,$$

where $p \in$ PROP and $a \in \mathbb{A}$. We will also use the following standard abbreviations, $\bot := \neg\top$, $\varphi \vee \psi := \neg(\neg\varphi \wedge \neg\psi)$, and $\hat{K}_a\varphi := \neg K_a\neg\varphi$.

On the semantic side, we take the most straightforward approach and use standard Kripke semantics with the only exception that the valuation function is a partial function. Formally:

Definition 1 (Partial possible world model (preliminary[3])) *A partial possible world model (or partial Kripke model) is a tuple* $\mathcal{M} = \langle W, (R_a)_{a \in \mathbb{A}}, V \rangle$, *where W is a non-empty set of worlds, R_a is a binary (equivalence) relation on W for each $a \in \mathbb{A}$, and V is a **partial** function from $W \times$ PROP into $\{0, 1\}$.*

For interpreting the language we specify a partial semantics based on two relations, namely a verification relation $\models$ and a falsification relation $=\!\!|$, as falsity is not the same as absence of truth in partial logics (Jaspars, 1994; Jaspars & Thijsse, 1996). The semantics, of what we will refer to as Partial Epistemic Logic (ParEL), is shown in Figure 1.

---
[3] We will change the definition slightly in the next section.

Modeling Truly Dynamic Epistemic Scenarios in PDEL

$\mathcal{M}, w \models p$	iff	$V(w,p) = 1$
$\mathcal{M}, w \models \top$		
$\mathcal{M}, w \models \neg\varphi$	iff	$\mathcal{M}, w =\!\!\mid \varphi$
$\mathcal{M}, w \models \varphi \wedge \psi$	iff	$\mathcal{M}, w \models \varphi$ and $\mathcal{M}, w \models \psi$
$\mathcal{M}, w \models K_a\varphi$	iff	$\forall v \in W : wR_a v$ implies $\mathcal{M}, v \models \varphi$
$\mathcal{M}, w =\!\!\mid p$	iff	$V(w,p) = 0$
$\mathcal{M}, w \not\models\!\!\mid \top$		
$\mathcal{M}, w =\!\!\mid \neg\varphi$	iff	$\mathcal{M}, w \models \varphi$
$\mathcal{M}, w =\!\!\mid \varphi \wedge \psi$	iff	$\mathcal{M}, w =\!\!\mid \varphi$ or $\mathcal{M}, w =\!\!\mid \psi$
$\mathcal{M}, w =\!\!\mid K_a\varphi$	iff	$\exists v \in W : wR_a v$ and $\mathcal{M}, v =\!\!\mid \varphi$

Figure 1: The semantics of Partial Epistemic Logic

If $\mathcal{M}, w \models \varphi$, we say that $\mathcal{M}$ verifies (or satisfies) φ at w, and if $\mathcal{M}, w =\!\!\mid \varphi$, we say that $\mathcal{M}$ falsifies φ at w. Moreover, if both $\mathcal{M}, w \not\models \varphi$ and $\mathcal{M}, w \not\models\!\!\mid \varphi$ then we say the φ is undefined at w in $\mathcal{M}$.

With the definition of $\varphi \vee \psi$ as $\neg(\neg\varphi \wedge \neg\psi)$, and $\hat{K}_a\varphi$ as $\neg K_a \neg\varphi$ the semantic clauses for disjunction and $\hat{K}_a$-modality become:

$\mathcal{M}, w \models \varphi \vee \psi$	iff	$\mathcal{M}, w \models \varphi$ or $\mathcal{M}, w \models \psi$
$\mathcal{M}, w =\!\!\mid \varphi \vee \psi$	iff	$\mathcal{M}, w =\!\!\mid \varphi$ and $\mathcal{M}, w =\!\!\mid \psi$
$\mathcal{M}, w \models \hat{K}_a\varphi$	iff	$\exists v \in W: wR_a v$ and $\mathcal{M}, v \models \varphi$
$\mathcal{M}, w =\!\!\mid \hat{K}_a\varphi$	iff	$\forall v \in W: wR_a v$ implies $\mathcal{M}, v =\!\!\mid \varphi$.

Note, nevertheless, that we do not always have $\mathcal{M}, w \models \varphi \vee \neg\varphi$, contrary to classical logic. Still, it is possible to translate ParEL into classical Epistemic Logic: It is possible to define two translations $(\cdot)^*$ and $(\cdot)_*$ on the set of formulas such that for every formula φ and every partial model $\mathcal{M}$, one can construct a classical model $\mathcal{M}^*$ such that:

$$\mathcal{M}, w \models \varphi \text{ iff } \mathcal{M}^*, w \models_c \varphi^*$$
$$\mathcal{M}, w =\!\!\mid \varphi \text{ iff } \mathcal{M}^*, w \models_c \varphi_*,$$

where the $\models_c$ is the classical consequence relation. For the details see (Langholm, 1988) and (Thijsse, 1992).

3 A partial version of Dynamic Epistemic Logic

We will now extend the ParEL of the previous section to a dynamic version. The resulting logic will be referred to as *Partial Dynamic Epistemic Logic*

(ParDEL). However, before we do this we will make a small change to the notion of a partial possible world model. We will not assume any set of agents or propositional variables given in advance, these will only be given in the model.

Definition 2 (Partial possible world model)   *A partial possible world model (or partial Kripke model) is a tuple* $\mathcal{M} = \langle W, A, P, (R_a)_{a \in \mathbb{A}}, V \rangle$, *where W is a non-empty set of worlds, A is a set of agents, P is a set of propositional variables, R_a is a binary (equivalence) relation on W for each $a \in \mathbb{A}$, and V is a partial function from $W \times P$ into $\{0, 1\}$.*

This will not affect the semantics given in Figure 1 with one small exception though. First note that if $p \notin P$, then neither $V(w, p) = 1$ nor $V(w, p) = 0$ is the case for any $w \in W$. Hence, if $p \notin P$, p will be undefined at all worlds $w \in W$, which is what we desire. However, for the knowledge operator we have to be a little more careful. What happens to the formula $K_a \varphi$ if $a \notin A$? If we take wR_av not to hold for any $w, v \in W$ if $a \notin A$, then we get the consequence that a will vacuously know everything. This we take to be undesirable, and it will affect our intuition about what happens when a enters the scenario and becomes part of the model. Thus, we will make the assumption that *"agents not in the model has no more knowledge than what is common knowledge to the agent in the scenario"*. This amounts to assuming that wR_av holds for all $w, v \in W$ and all $a \notin A$. With this convention the semantic clause for K_a in Figure 1 will not need to change. One way of viewing this change is that we do not only allow for partiality of atomic proposition, but we also allow for "partiality of agents".

The language build from the set of agents A and the set of propositional variables P, in the recursive way of the previous section, will be denoted by $\mathcal{L}(A, P)$.

With this small change to ParEL, we can now move to a dynamic version of the logic by introducing event models. We will include post-conditions in our event models, since we want to be able to introduce new propositional variables into a model and specify their truth values. Without postconditions, standard event models of DEL cannot "change" propositional variables and we would therefore need to include them and their truth values already in the beginning of the modeling process—which is exactly what we what to avoid! Though, before we can define event models we need the notion of a substitution.

Definition 3 (Substitution)   *A substitution is a function* $\sigma \colon X \to \mathcal{L}$, *where X is a finite set of propositional variables and $\mathcal{L}$ is a language. X will* be called the domain of σ and will also be denoted $dom(\sigma)$. For a fixed language $\mathcal{L}$ and a fixed set of propositional variables P, the set of all substitutions $\sigma\colon X \to \mathcal{L}$, with $X \subseteq P$, will be denoted by $sub(P, \mathcal{L})$.

We can now give the following definition of an event model:

Definition 4 ((Partial) event model) *A (partial) event model is a tuple $\mathcal{E} = \langle E, B, Q, (S_a)_{a \in B}, \mathsf{pre}, \mathsf{post}\rangle$, such that E is a non-empty finite set of events, B is a set of agents, Q is a set of propositional variables, S_a is a binary (equivalence) relation on E for each $a \in B$, $\mathsf{pre}\colon E \to \mathcal{L}(B,Q)$ is a precondition function assigning a precondition to each event, and $\mathsf{post}\colon E \to sub(Q, \mathcal{L}(B,Q))$ is a postcondition function specifying what propositional variables will change if a given event happens.*

In what follows it will be convenient to let $\mathsf{post}(e)(p)$ refer to p whenever $p \in Q \setminus dom(\mathsf{post}(e))$ and to $\bot$ whenever $p \notin Q$, thus, we will make this convention. With this and the notion of an event model we can now specify how a partial possible world model changes if an event model "happens". As is usual in DEL, we do this through a notion of "product update":

Definition 5 (Product update) *Let $\mathcal{M} = \langle W, A, P, (R_a)_{a \in \mathbb{A}}, V\rangle$ be a partial possible world model and $\mathcal{E} = \langle E, B, Q, (S_a)_{a \in B}, \mathsf{pre}, \mathsf{post}\rangle$ an event model. The product update $\mathcal{M} \otimes \mathcal{E} = \langle W', A', P', (R'_a)_{a \in \mathbb{A}}, V'\rangle$ is defined by:*

- $W' = \{(w, e) \in W \times E \mid \mathcal{M}, w \models \mathsf{pre}(e)\}$
- $A' = B$
- $P' = Q$
- $(w, e) R'_a (v, f)$ *iff*
  - $w R_a v$ *and* $e S_a f$, *if* $a \in B \cap A$, *and*
  - $e S_a f$, *if* $a \in B \setminus A$
- $V'((w,e), p) =$
  - 1, *if* $p \in Q$, *and* $\mathcal{M}, w \models \mathsf{post}(e)(p)$
  - 0, *if* $p \in Q$, *and* $\mathcal{M}, w \not\models \mathsf{post}(e)(p)$

We now add constructs of the form $[\mathcal{E}, e]\varphi$ to the language for every event model $\mathcal{E}$, event e of $\mathcal{E}$, and every formula φ. The semantics of the formula $[\mathcal{E}, e]\varphi$ is given in Figure 2 and the resulting logic will be referred to as *Partial Dynamic Epistemic Logic* (ParDEL).[4]

---
[4]There is at least one alternative definition of the semantics for $[\mathcal{E}, e]\varphi$, for instance we

$\mathcal{M}, w \models [\mathcal{E}, e]\varphi$	iff	$\mathcal{M}, w \models \text{pre}(e)$ implies $\mathcal{M} \otimes \mathcal{E}, (w, e) \models \varphi$
$\mathcal{M}, w \dashv [\mathcal{E}, e]\varphi$	iff	$\mathcal{M}, w \models \text{pre}(e)$ and $\mathcal{M} \otimes \mathcal{E}, (w, e) \dashv \varphi$

Figure 2: The semantics of the event modality

This notion of an event model is a generalization of the standard definition of an event model (with post-conditions) in DEL. Thus all kinds of dynamics that can be represented in DEL can be represented in ParDEL. Additionally, with this new notion of an event model we can also represent the entering of new agents and atomic facts. To see this, let us return to the example in the introduction.

First of all, it is fairly obvious that the initial situation of the example can be captured by a possible world model, where the set of possible worlds are the set of possible deals of cards, and two worlds are indistinguishable for an agent if he holds the same cards in both worlds. Propositional variables for each agent holding a particular card can be introduced into the language allowing us to talk about the scenario. The event that Chuck puts three cards face-up on the table or that Arnold shows one of his cards to Bruce, are also easily describable in classical DEL. Nevertheless, the entering of a new agent, such as Bruce's wife, cannot be modeled in classical DEL.

Let us see how to model the entering of Bruce's wife in ParDEL. First assume that Bruce's wife (d) enters completely ignorant of any of the player's cards. This is captured by applying the following partial event model $\mathcal{E} = \langle E, B, Q, (S_a)_{a \in B}, \text{pre}, \text{post} \rangle$, where

- $E = \{e\}$, $B = A \cup \{d\}$, $Q = P$
- $S_a = S_b = S_c = S_d = \{(e, e)\}$
- $\text{pre}(e) = \top$, $\text{post}(e) = \emptyset$,

to the initial situation. If Bruce's wife enter seeing Chuck's cards the partial event model would be defined by:

---

could define that

$$\mathcal{M}, w \models [\mathcal{E}, e]\varphi \text{ iff } \mathcal{M}, w \dashv \text{pre}(e) \text{ or } \mathcal{M} \otimes \mathcal{E}, (w, e) \models \varphi,$$

which is not equivalent to the definition given in Figure 2. At hand, the author is not aware of any "knock-down" arguments for the choice of semantics given by Figure 2, except that it seems intuitive, it resembles the semantics for classical DEL, and it allows for nice reduction equivalences, as discussed in Section 4.

- $E = C \times C$ (where C is the set of all cards), $B = A \cup \{d\}$, $Q = P$
- $S_a = S_b = E \times E$, $S_c = S_d = \{(e,e) \mid e \in E\}$
- $\mathsf{pre}((c_1, c_2)) =$ "c holds card c_1 and c_2", $\mathsf{post}(e) = \emptyset$.

Now, consider the case if *"Chuck announces that he has the ability to see through cards, but he never uses it unless he has an incredibly bad hand"*. First of all, we need to agree on what a bad hand is. However, whatever cards constitute a bad hand, that Chuck holds a bad hand can be completely defined in the language by a formula φ_{cbh} that is a disjunction of all the possible bad hands Chuck can possess. Secondly, we need a new propositional variable q for "Chuck has looked at Arnold's and Bruce's hands". With this in place, the event can now be described by a partial event model $\mathcal{E} = \langle E, B, Q, (S_a)_{a \in B}, \mathsf{pre}, \mathsf{post} \rangle$, where

- $E = \{e, f\}$, $B = A$, $Q = P \cup \{q\}$
- $S_a = S_b = E \times E$, $S_c = \{(e,e), (f,f)\}$
- $\mathsf{pre}(e) = \varphi_{cbh}$, $\mathsf{pre}(f) = \neg \varphi_{cbh}$, $\mathsf{post}(e) = q$, $\mathsf{post}(f) = \neg q$.

Note that, in this way, it is ensured that Chuck has seen the cards of Arnold and Bruce if, and only if, he has a bad hand ($q \leftrightarrow \varphi_{cbh}$), since φ is a Boolean formula where none of the propositional variables are in $dom(\mathsf{post})$.

These small examples hopefully give a hint as to how general the event models of ParDEL are when it comes to model the epistemic aspects of the entering of new agents and new atomic facts. Thus, ParDEL is an excellent candidate for the framework asked for in the introduction.

4 Reduction laws for Partial Dynamic Epistemic Logic

In classical DEL it turns out that the event modalities $[\mathcal{E}, e]$ are eliminable, in the sense that every formula of DEL can be translated, in a truth-preserving way, into a formula without any event modalities, i.e. a formula of classical Epistemic Logic. This is due to the validity of so called *reduction axioms*. The reduction axioms of DEL are (van Ditmarsch & Kooi, 2008):

$$
\begin{aligned}
{[\mathcal{E}, e](\varphi \wedge \psi)} &\leftrightarrow [\mathcal{E}, e]\varphi \wedge [\mathcal{E}, e]\psi \\
{[\mathcal{E}, e]\neg\varphi} &\leftrightarrow \mathsf{pre}(e) \to \neg[\mathcal{E}, e]\varphi \\
{[\mathcal{E}, e]K_a\varphi} &\leftrightarrow \mathsf{pre}(e) \to \bigwedge_{f \in E, eS_a f} K_a[\mathcal{E}, f]\varphi \\
{[\mathcal{E}, e]p} &\leftrightarrow \mathsf{pre}(e) \to \mathsf{post}(e)(p),
\end{aligned}
$$

$\mathcal{M}, w \models [\mathcal{E}, e]p$	iff	$\mathcal{M}, w \models \mathsf{pre}(e) \to \mathsf{post}(e)(p)$
$\mathcal{M}, w \models [\mathcal{E}, e]\top$		
$\mathcal{M}, w \models [\mathcal{E}, e]\neg\varphi$	iff	$\mathcal{M}, w \models \mathsf{pre}(e) \to \neg[\mathcal{E}, e]\varphi$
$\mathcal{M}, w \models [\mathcal{E}, e](\varphi \wedge \psi)$	iff	$\mathcal{M}, w \models [\mathcal{E}, e]\varphi \wedge [\mathcal{E}, e]\psi$
$\mathcal{M}, w \models [\mathcal{E}, e]K_a\varphi$	iff	$\mathcal{M}, w \models \mathsf{pre}(e) \to \bigwedge_{f \in E, eS_a f} K_a[\mathcal{E}, f]\varphi$ [1]
$\mathcal{M}, w =\!\!\!| [\mathcal{E}, e]p$	iff	$\mathcal{M}, w =\!\!\!| \neg\mathsf{pre}(e) \vee \mathsf{post}(e)(p)$
$\mathcal{M}, w \not\!\!\dashv [\mathcal{E}, e]\top$		
$\mathcal{M}, w =\!\!\!| [\mathcal{E}, e]\neg\varphi$	iff	$\mathcal{M}, w =\!\!\!| \neg\mathsf{pre}(e) \vee \neg[\mathcal{E}, e]\varphi$
$\mathcal{M}, w =\!\!\!| [\mathcal{E}, e](\varphi \wedge \psi)$	iff	$\mathcal{M}, w =\!\!\!| [\mathcal{E}, e]\varphi \wedge [\mathcal{E}, e]\psi$
$\mathcal{M}, w =\!\!\!| [\mathcal{E}, e]K_a\varphi$	iff	$\mathcal{M}, w =\!\!\!| \neg\mathsf{pre}(e) \vee \bigwedge_{f \in E, eS_a f} K_a[\mathcal{E}, f]\varphi$ [1]

[1] Here it is assumed that $a \in A \cap B$ of $\mathcal{M}$ and $\mathcal{E}$.

Figure 3: The reduction laws of ParDEL

where $[\mathcal{E}, e]$ denotes an event modality of classical DEL. From these validities it is not hard to see that a formula can be translated into a formula without event modalities: Since the reduction axioms allow us to reduce the complexity of a formula within the scope of an event modality until we reach propositional variables at which the last reduction axiom allow us to completely do away with the event modality. For this reason, adding the reduction axioms to an axiomatization of Epistemic Logic result in a sound and complete axiomatization of DEL.

When moving to ParDEL the question is: is a similar trick possible? And the answer is: in some sense yes. We do not obtain a single truth-preserving translation of ParDEL into ParEL, however we do obtain *reduction laws*, as shown in Figure 3[5], that allow us to successively reduce the complexity within the scope of an event modality $[\mathcal{E}, e]$.

There is a small problem with the reduction laws shown in Figure 3. On the right-hand side of the reduction laws in the top half of the figure, we used the implication connective "$\to$". However, for the reduction laws to work, we cannot take $\varphi \to \psi$ to be $\neg\varphi \vee \psi$, since we need $\to$ to satisfy:

$$\mathcal{M}, w \models \varphi \to \psi \text{ iff } \mathcal{M}, w \models \varphi \text{ implies } \mathcal{M}, w \models \psi,$$

which $\neg\varphi \vee \psi$ does not. In fact, the implication we need cannot be defined

---

[5] The reduction laws only look like in Figure 3 if $a \in A \cap B$ of $\mathcal{M}$ and $\mathcal{E}$. If $a \notin A$, then the K_a operator on the right side should be replaced by a global modality. If additionally, $a \notin B$, then the large conjunction on the right should be over all $f \in E$.

in our present language. This is a well-known issue in partial logic as the connectives $\neg$ and $\wedge$ are not functionally complete in partial logic (Thijsse, 1992). Instead of adding the implication to our language we will follow the common route and add a classical negation $\sim$. Hence, we add the additional syntactic clause that whenever φ is a formula, so is $\sim\varphi$. On the semantic side we add the following semantic clauses:

$$\mathcal{M}, w \models \sim\varphi \text{ iff } \mathcal{M}, w \not\models \varphi$$
$$\mathcal{M}, w =| \sim\varphi \text{ iff } \mathcal{M}, w \models \varphi.$$

The logic obtained by adding $\sim$ to ParDEL will be denoted ParDEL($\sim$) and similar ParEL($\sim$) is ParEL with $\sim$ included. With the new negation, we can define $\varphi \to \psi$ as $\sim\varphi \vee \psi$. Reduction laws for the new negation $\sim$ are also obtainable, in the sense that

$$\mathcal{M}, w \models [\mathcal{E}, e]\sim\varphi \text{ iff } \mathcal{M}, w \models \mathsf{pre}(e) \to \sim[\mathcal{E}, e]\varphi$$
$$\mathcal{M}, w =| [\mathcal{E}, e]\sim\varphi \text{ iff } \mathcal{M}, w =| \neg\mathsf{pre}(e) \vee \sim[\mathcal{E}, e]\varphi$$
$$\text{(or equivalently } \mathcal{M}, w =| \neg\mathsf{pre}(e) \vee \neg[\mathcal{E}, e]\varphi).$$

It is now possible to define two translations t^+ and t^- of ParDEL($\sim$) into ParEL($\sim$) such that for every formula φ of ParDEL($\sim$):

$$\mathcal{M}, w \models \varphi \text{ iff } \mathcal{M}, w \models t^+(\varphi)$$
$$\mathcal{M}, w =| \varphi \text{ iff } \mathcal{M}, w \models t^-(\varphi).$$

We will leave the details of the translations as well as the proof of this claim for future research. Moreover, using the reduction laws for providing a proof system for ParDEL is left for future research, as well. Finally, given the translations of ParEL into classical Epistemic Logic, mentioned in Section 2, it should be possible to translate questions of verification and falsification of ParDEL formulas into questions of satisfiability in classical Epistemic Logic. Again we leave the details for future research.

5 Conclusion and future research

In this paper, it was argued that classical DEL is not powerful enough to model all kinds of natural dynamics, such as entering of new agents or facts, and that from a modeling perspective a partial version of DEL seems natural and might make "the art of modeling" more into a systematic science. Moreover, ParDEL can be viewed as an alternative approach to modeling awareness (Thijsse, 1992) and its dynamics. In addition to this, it should be mentioned that negative introspection also becomes more natural in ParDEL, as expressed by van der Hoek, Jaspars, and Thijsse (1996, p. 323): *"if an agent considers no world possible in which p is true, it may simply mean that he does not reckon with p, rather than that he knows that $\neg p$ is true"*. In this way, several sources of logic omniscience are eliminated. Finally, ParDEL may provide for a more "economical" representation of ignorance/unawareness. For instance, if a single agent is ignorant/unaware about five atomic facts this requires a possible world model with 32 worlds in classical Epistemic Logic, whereas in a partial version of Epistemic Logic one can simply start with a possible world model consisting of a single world and only expand the world if the agent starts to consider some of the five atomic facts. In conclusion, there are numerous reasons for considering a partial version of DEL, as done in this paper.

Even though the apparent defects of classical DEL are highlighted, this paper draws heavily on inspiration from the DEL framework. Actually, this paper can be seen as a demonstration of the great generality and potentials of the DEL framework. For instance, it was demonstrated that event models (in a slightly generalized form) are a very natural framework for dealing with evolution/extensions of partial possible world models.

Inspired by the event models of DEL, another version of partial DEL might seem worthwhile investigating. In partial possible world models, one could equip worlds with *sets* of classical valuations instead of one partial valuation. This will make classical event models (with Boolean preconditions) epistemic models as well. Moreover, the product update operation becomes commutative. The investigation of this alternative version of partial DEL is left for future research.

Finally, several technical issues relating to ParDEL remain for future research as discussed at the end of Section 4.

References

Areces, C., & ten Cate, B. (2007). Hybrid Logics. In P. Blackburn, J. van Benthem, & F. Wolter (Eds.), *Handbook of Modal Logic* (pp. 821–868). Amsterdam: Elsevier.

Blackburn, P. (2000). Representation, Reasoning, and Relational Structures: A Hybrid Logic Manifesto. *Logic Journal of IGPL, 8*, 339–365.

Fagin, R., & Halpern, J. Y. (1988). Belief, Awareness, and Limited Reasoning. *Artificial Intelligence*, *34*, 39–76.

Jaspars, J. (1994). *Calculi for Constructive Communication, A Study of the Dynamics of Partial States*. Tilburg, Amsterdam: ITK/ILLC Publications. (ITK/ILLC Dissertation Series)

Jaspars, J., & Thijsse, E. (1996). Fundamentals of Partial Modal Logic. In P. Doherty (Ed.), *Partiality, Modality, and Nonmonotonicity* (pp. 111–141). Stanford: CSLI Publications.

Langholm, T. (1988). *Partiality, Truth, and Persistence* (Vol. 15). Stanford: CSLI Publications.

Thijsse, E. (1992). *Partial Logic and Knowledge Representation*. PhD Thesis, Tilburg University.

van Benthem, J. (2009). *The Art of Modeling* (Tech. Rep. No. PP-2009-06). Amsterdam: ILLC.

van der Hoek, W., Jaspars, J., & Thijsse, E. (1996). Honesty in Partial Logic. *Studia Logica*, *56*, 323–360.

van Ditmarsch, H., & French, T. (2009). Awareness and Forgetting of Facts and Agents. In *Web Intelligence/IAT Workshops* (pp. 478–483). Dagstuhl: IEEE.

van Ditmarsch, H., & French, T. (2011). Becoming Aware of Propositional Variables. In M. Banerjee & A. Seth (Eds.), *Proceedings of the 4th Indian Conference on Logic and Its Applications (ICLA 2011)* (pp. 204–218). Berlin, Heidelberg: Springer.

van Ditmarsch, H., & Kooi, B. (2008). Semantic results for Ontic and Epistemic Change. In G. Bonanno, W. van der Hoek, & M. Wooldridge (Eds.), *Logic and the Foundations of Game and Decision Theory (LOFT 7)* (pp. 87–117). Amsterdam: Amsterdam University Press.

van Ditmarsch, H., van der Hoek, W., & Kooi, B. (2008). *Dynamic Epistemic Logic*. Dordrecht: Springer.

Jens Ulrik Hansen
Lund University
Sweden
E-mail: `jens_ulrik.hansen@fil.lu.se`

First-Person Logical Theories and Third-Person Logical Theories

John T. Kearns

Abstract: A first-person account, a description, explanation, or theory, is an account of experience or actions from the perspective of the person having the experience or performing the actions. A first-person account is really for that first person, and can only be evaluated from the perspective of the person whose account it is. A third-person account is the same for everyone, and can be evaluated from a neutral, or impersonal standpoint. Certain ethical theories and logical theories have an essentially first-person character. This does not make them capricious or subjective, for a first-person theory can be governed by objective principles which are the same for everyone. The theories of illocutionary logic, the logic of speech acts, are essentially first-person theories. These theories explore, both semantically and deductively, the commitments generated by performing assertive illocutionary acts. The present paper argues that even standard logical theories are first-person theories. They have, superficially, the appearance of third-person theories, but they are actually truncated and abstract versions of explicitly first-person theories.

Keywords: philosophical logic, illocutionary logic, speech acts

1 Two kinds of theory

We can distinguish two kinds of description, or explanation, or theory of something. Let us call descriptions, explanations, and theories *accounts*. A third-person account is the same for all people who know or accept the account, and even for all people who know about the account. The Ptolemaic theory of the Earth, Sun, planets and fixed stars is a third-person theory. It says how the various heavenly bodies are located, and how they move with respect to one another. The theory doesn't reflect any particular person's perspective.

A *first-person account* of something reflects or incorporates a particular person's perspective on that something. It is an account which only that person could give or adopt, or even consider giving. The account treats something that only the particular person can experience or access, or that only she can access in a way that is relevant to the account. Each person has her own experience. Each person's account of her own perceptual experience, from the "inside" as it were, is essentially first-person. Such an account might be concerned with the objects perceived and their perceived properties and relations, with interpreting her perceptions, or with how she does or can direct her attention to various objects, events, and features. However, while each person's account is essentially her own, certain principles may govern, or apply to, or be true of everyone's first-person account. For everyone's experience, for example, nothing is or can be red all over and green all over.

Just as no one can have, or share, someone else's experience, so each person's actions are her own. One person can act on behalf of another, but no person can perform another person's actions. There might be a third-person account of people and their behaviors "from the outside," but each person has her own perspective on her own acts, and this enables her to give a first-person account of her own acts and actions. One couldn't even give a third-person account of acts and actions, if it were concerned with more than observable behavior, without borrowing elements from her own first-person understanding.

A first-person account of one's own actions might be historical in the sense that it is concerned with what one has done and tried to do, and with her successes and failures. A first-person theory of action might also be normative, incorporating principles about what are the right, or correct, things to do, and with how one must, or ought to, carry out certain acts and activities. A normative theory of this sort can incorporate principles which are the same for everyone, but indicate how the first person involved must, or should, implement the principles in her own particular circumstances. In discussing G.E. Moore and ethics, Smith (2003) has explained how different people can each have their own first-person ethical theories, while each such theory incorporates a sub-theory that has a somewhat third-person "look." These first-person theories might recommend actions that favor the first person at the expense of others, but still accept some principles that each person uses to determine the same positive and negative evaluations for a limited range of situations. The sub-theories of the first-person theories are only "apparently" third-person theories.

I think that standard logical theories are like these ethical sub-theories in having an apparently third-person character, while actually belonging to larger, essentially first-person, theories. Logical theories are, I think, best understood to be theories of *speech acts*, or *language acts*. These are meaningful acts performed with expressions. Both the people who produce the expressions that are used, and those who read expressions or understand spoken expressions perform language acts, but it is common to focus on acts in which expressions are both produced and used, and I shall do that here. One of these acts can be performed by speaking, or writing, or by merely thinking with words. A *sentential act* is an act performed with a sentence of a natural language. I shall understand a *statement* to be a sentential act that can appropriately be evaluated in terms of truth and falsity. On this understanding, someone who makes a statement need not accept that statement.

A sentential act can be performed with a certain (*illocutionary*) force to constitute an *illocutionary act* like an assertion, a promise, or an apology. Statements themselves can be used with different forces to constitute *assertions*, *denials*, and positive and negative *suppositions*—these are examples of *assertive illocutionary acts*. An assertion is here taken to be an act of producing a statement and accepting that statement as being or representing what is the case, or else an act of producing a statement and reaffirming one's continued acceptance of the statement. While many illocutionary acts require an addressee, an assertion, as I understand it, does not. A denial is an act of ruling out the acceptance of a statement, because of the statement's failure to fit the world. A positive supposition is an act of temporarily accepting a statement, to explore the consequences of doing so, while a negative supposition is an act of impeding its temporary acceptance. Suppositions play an important role in many deductive arguments, and are often discharged or canceled in the course of an argument.

2 Logical theories

I understand a logical theory to consist of a *framework*, or *frame*, containing a specialized formal language, often artificial, a semantic account for that language, and a deductive system for establishing results concerning expressions of the language, together with derivations in the deductive system, as well as the results established by these derivations. This is a logical theory, *narrowly conceived*. Conceived more broadly, the theory also includes theorems or metatheorems establishing results about the language, semantic account, and deductive system. Expressions of the logical language represent language acts that either are or might be performed with the languages we actually speak, and the semantic account in the logical framework is for these language acts.

Standard logical theories are concerned with statements and semantic features of statements. The fundamental theories are classical theories, which *codify* logical truth and implication or logical consequence. Nonclassical theories often codify other features than (ordinary) logical truth and deal with limited consequence relations. Since standard theories study language acts, and two people can't perform the same language acts, standard theories must be first-person theories. However, there are statements made by different people which are essentially the same as, or essentially similar to, one another. Standard logical theories are concerned with statements of this kind, and, as a result, standard logical theories have a third-person "look."

In developing standard logical theories, it is common to abstract away from the persons making the statements that are studied, for it is convenient to regard different people as making the same statements. It is also common to abstract away from the fact that speech acts are the objects being studied, and to consider those objects to have the more splendid natures often attributed to propositions.

The semantic properties explored in standard theories are based on truth conditions of statements. I think the study of statements and their truth conditions is best regarded as ontological, or *ontic*. Statements in a perspicuous logical language represent the ontological "constitution" of reality, and truth conditions are formulated in terms of the objects "answering" to referring and denoting acts.

A logical theory for those language acts that can't be performed by different people, and that, when performed by one person, are not essentially similar to acts performed by someone else, must be an *explicitly and evidently* first-person theory. Illocutionary acts are acts of this kind. For statements, the important semantic features involve truth conditions. Two people who each make statements that are essentially similar will make statements which have the same truth conditions. For *assertive* illocutionary acts, like assertions, denials, and suppositions, the important semantic feature is *rational commitment*.

Statements are true or false (and sometimes, perhaps, neither one), statements imply and are incompatible with one another. Assertive illocutionary acts are not true or false, although their component statements are (usually) one or the other, and assertive illocutionary acts are not linked by implication or incompatibility. But a person who performs some assertive illocutionary acts is committed by doing so to perform further acts, though her acts have no such effect on other people, not even on people who recognize and understand her acts. Commitment is an *epistemic* feature of assertive illocutionary acts and of the people who perform them.

Rational commitment is a commitment to do or not do something, or to continue in a certain state like that of accepting a given statement. Deciding to perform an action commits a person to perform that action. Some commitments are conditional while others are not. My commitment to close the upstairs windows if it rains while I am at home is conditional, but decisions can establish unconditional commitments.

Performing some acts often commits a person to perform others. Accepting or asserting a given statement will commit a person to accept other statements. These commitments are conditional on the person giving the matter some thought, and having an interest in them. Asserting that today is Wednesday commits me to accept the statement that either today is Wednesday or it is now snowing in Beijing. But I have no interest in that further statement, and would not normally consider it. Still, it would be irrational to accept that today is Wednesday and refuse to grant that either today is Wednesday or it is now snowing in Beijing.

The truth conditions of statements are *ontic* semantic features of statements. Semantic features defined in terms of truth conditions are also ontic features. The truth of a statement doesn't depend on its being known by anybody, and some statements imply another statement or not, regardless of whether anyone recognizes this, even regardless of whether anyone *can* recognize it. There are simple cases of implication that everyone can recognize, and complicated cases which are difficult to recognize, but the complicated cases are not built up in some way from simple cases.

In contrast, rational commitment is an *epistemic* feature. It depends on being recognized, or, at least, on people being able to recognize it. But some qualification is in order here. We need to distinguish *immediate* from *mediate*, or remote, commitment. If doing X *immediately* commits a given person to doing Y, then once that person does X, she must be able to recognize (if she thinks about it at all) that she is committed to doing Y. But if doing X_1 immediately commits a person to doing X_2, doing X_2 immediately commits her to doing $X_3,\ldots$, and doing X_{n-1} immediately commits her to doing X_n, then doing X_1 may only *mediately* commit her to doing X_n. Mediate commitment may not be evident to the person who is committed. It is immediate commitment, not mediate or remote commitment, which motivates a person to act.

The distinction between mediate and immediate commitment is indicative of the temporal character, or the temporal dimension, of commitment, in contrast with truth, and truth-conditional features like entailment and implication, which don't have such a character. A person's commitments lead him to act or refrain from acting in the future, or to remain in a certain state (also in the future). The implications of a set of statements don't call for any kind of behavior.

A given person, at a time, is characterized by her commitments to perform or not perform certain acts, including her commitments to accept or deny certain statements. In a genuine deduction, a *natural* deduction, a person traces the immediate commitments of her illocutionary acts and of the states that her acts reflect. A truly natural deduction involves moves from illocutionary acts to illocutionary acts, and the correctness of such a deduction depends on both truth conditions and illocutionary force. But truth conditions by themselves are *inert* when it comes to constructing deductive arguments. A person needs to *recognize* the truth conditions of statements, and to recognize relations defined in terms of truth conditions, in order for the truth conditions to impact her argument. Her recognizing the relations can generate a commitment to perform an illocutionary act. She must also recognize her commitments before she can act upon them, but her immediate commitments are evident to her once she attends to them. Immediate commitment is the "engine" which drives the rational process of deduction.

3 Illocutionary logic

Illocutionary logic was introduced by Vanderveken and Searle (1985), but I have provided an alternative formulation in a number of places, including (Kearns, 2000, 2006, 2007). Illocutionary logic is the logic of speech acts, or language acts. Full theories of illocutionary logic are explicitly and evidently first-person theories. The frame of a standard logical language contains non-logical expressions, some distinctively logical expressions like connectives and quantifiers, and sentences composed from them. The language is "interpreted" by functions with respect to a domain of objects or with respect to a whole universe of worlds. And there is a deductive system for tracing or tracking logical consequence. The frame for a full theory of illocutionary logic is obtained from the frame of a standard theory by adding things to it.

Sentences in the standard theory's logical language represent statements. The standard theory's language is enriched with various first-person expressions to obtain the language of the full illocutionary theory. Most prominent, perhaps, are the *illocutionary operators* with which a speaker makes explicit the force of her own assertive acts. These are prefixed to the (*plain*) sentences which represent statements to obtain *completed sentences*, which represent illocutionary acts being performed by the speaker. The basic illocutionary operators are the following:

⊢ — the sign of assertion

⊣ — the sign of denial

∟ — the sign of positive supposition

¬ — the sign of negative supposition

Suppositions are acts of temporarily accepting or rejecting statements. If a conclusion is deduced from premises which include suppositions, that conclusion is also a supposition.

Genuine arguments, arguments in real life as opposed to those in logic books and logic courses, are *speech-act arguments*. The premises are illocutionary acts and so are the conclusions. The deductive system in a fully developed illocutionary theory enables a person to construct and study speech act arguments, but she is studying these arguments from her own perspective. They are arguments from her own assertions, denials, and suppositions to conclusions which she herself asserts, denies, or supposes.

4 Illocutionary semantics

A logical theory which deals with assertive illocutionary acts must be a first-person theory. Each person's theory will contain illocutionary acts (completed sentences) which record her own knowledge and belief (as well as disbelief). In developing her illocutionary theory, each person will enlarge her (explicit) knowledge and belief. If each person has her own theory, and is responsible for developing her own theory, it may seem that there is no work for a professional logician to carry out. But, in fact, the logician's job is to uncover principles which must govern, or apply to, each person's theory. The logician can develop a *generic* version of a first-person theory. Each person can use this generic theory in developing her own distinctive theory.

What are common to different first-person theories of assertive illocutionary acts are deductive and semantic principles. The deductive principles are for constructing the arguments that actually figure in "real-life."

(Real-life arguments don't have statements as premises and statements as conclusions, instead they begin with assertions, denials, or suppositions, and conclude with acts of these kinds.) The illocutionary operators in systems of illocutionary logic are for making explicit the illocutionary force with which statements are "performed." These are first-person expressions, just as the pronouns 'I,' 'my,' and 'me' are first-person expressions, and these operators are indexical to the person who uses them. The illocutionary operators are not used to perform referring acts, and they are not used to characterize the statements to which they are prefixed. Instead a person uses them to indicate or make explicit what she is doing. For example, the language user can write '$\vdash A$' to show that she is asserting A, or '$\dashv A$' to make her denial explicit.

Deductive principles formulated for explicitly first-person logical theories accommodate both truth conditions and illocutionary force, for these are what determine the commitments of a person's assertive illocutionary acts. The semantic accounts for explicitly first-person theories also have a distinctive character. To deal with the truth conditions of statements, we employ functions which assign things to expressions, and this determines which statements are true and which aren't. When we consider a statement that different people can make, its truth conditions are independent of whose statement it is. The commitments of a particular person's assertive illocutionary acts depend on that person, and on what she knows or believes.

The semantic account for an explicitly first-person logical theory is formulated for a particular person at a particular time. At that time, the person will be committed by her knowledge and belief, including her knowledge of the language, to assert some statements and deny others. With respect to a person at a time, the plus sign '+' is used for the value *committed to perform*. *Commitment valuations* assign this value to completed sentences of the illocutionary language in ways we can specify, and this enables us to determine what it takes for a speech act argument to be *deductively correct*, and further enables us to design deductive systems which can be shown to be sound and complete.

The semantic account is *momentary* in the following sense. For person P and time t, the *moment determined by P at t* is the period that begins with t and continues so long as P makes only assertions or denials that she is committed to make by her knowledge or belief at t. Once she makes a "new" assertion or denial, perhaps on the basis of experience, the moment is ended. Deductive arguments trace a person's momentary commitments, so soundness and completeness proofs for an explicitly first-person theory

1st and 3rd Person Logical Theories

will establish that the momentary arguments sanctioned by the deductive system are adequate to the momentary commitments of particular persons at particular times.

Because logic is a normative subject, the (generic) semantic account is provided for an idealized language user whom I call the *designated subject*, for whom I use feminine pronouns. The account designed for this person allows us to determine which arguments are correct for everyone who has the relevant beliefs and disbeliefs, and who (perfectly) understands the language she is speaking.

Each person has her own illocutionary logical theory, which incorporates her own explicit knowledge and belief. Although each person's theory is her own, the rest of us can judge the adequacy of her arguments, once we are informed what it is she knows and believes. Even standard logical theories are first-person theories, because they are concerned with speech acts, which are always performed by particular people. In standard theories, we can safely ignore which people these are, but we can't get them out of the picture. And standard theories are concerned with truth-conditional features which don't differ from one time to another. But standard logical theories are really truncated versions of full theories of illocutionary logic.

5 Additional first-person logical theories

It isn't simply assertive illocutionary acts which call for explicitly first-person logical theories. Statements made with first-person expressions like 'I,' 'me,' 'my,' etc. are also essentially first person, and call for an explicitly first-person logical theory. Indeed, such theories are required in order to understand and accommodate a number of linguistic phenomena and linguistic puzzles. One of the simplest of these is Moore's Paradox. This was originally discussed with respect to a speaker who asserts the following statement:

> It's raining, but I don't believe it.

Moore noted that the statement has an inconsistent "feel" to it, even though the statement is consistent. He wondered what it is that is wrong with the statement.

To understand what is wrong, we must distinguish semantic features of statements from semantic features of illocutionary acts. A statement is (*semantically*) *inconsistent* if the truth conditions of the statement cannot be satisfied, and a set of statements is *inconsistent* if the truth conditions of the sentences in the set cannot be simultaneously satisfied. It is *incoherent* for a person to both assert and deny a single statement. Assertive illocutionary acts that commit a person to perform incoherent acts are themselves *incoherent*. For whoever is the I in question, the statement:

> It's raining, but I don't believe it.

will be consistent. It often rains when I'm not aware that it is raining, and this is also true for everyone else. But while such a statement will be consistent for whoever makes it, the statement cannot coherently be accepted, or asserted. Anyone who asserts a statement A will be committed to accept "I believe that A." But then anyone who asserts her own first-person version of the statement above will be committed to assert both that she believes that it is raining and that she doesn't believe that it is raining. She will be committed to assert an inconsistent statement, and doing this is incoherent.

An explicitly first-person logical theory is required to understand and explain features of one's own assertive illocutionary acts, though it is often sufficient to develop the generic version of such a theory. Simple assertions, denials, and suppositions do not exhaust the assertive acts that need explaining. In English, for example, indicative conditional sentences are often used to make conditional assertions, denials, and suppositions. In fact, there are conditional illocutionary acts of various kinds. There are conditional promises, conditional requests, conditional orders, and even conditional apologies, as illustrated by the following:

> If you must work late, I promise to pick you up when you are finished.
> If you can afford it, will you please lend me ten dollars.
> If tomorrow is the day that garbage is collected, then put the garbage can out by the street.
> If I stepped on your foot, I apologize.

These conditional acts do not contain or otherwise involve conditional *statements*. They are acts that "take effect" when the relevant person realizes that the appropriate condition "obtains." A conditional assertion, like this one:

> If Frank studied last night, he will surely get an A on the test today.

commits the speaker to accept that Frank will get an A on the test once she learns (if she does) that Frank studied last night, while a denial:

> If Frank studied last night, then he won't do badly on the test today.

commits the speaker to deny that Frank will do badly, on the same condition. Conditional assertions and denials aren't statements and don't have truth conditions. They need to be explained and explored by developing a first-person logical theory.

Epistemic modal logic is an essentially first-person subject matter, and suitable theories are essentially first-person theories. On no reasonable construal of knowledge or belief does a person know the consequences of everything she knows, or believe the consequences of everything she believes. But a person is committed by what she knows and believes to accept many statements and to deny many others. If '$\Box^K A$' has the significance *I am committed by my current knowledge to accept A*, and '$\Box^B A$' has the significance *I am committed by my current (justified) beliefs to accept A*, then the following inference principles are evidently correct when, first, the premiss assertion has the force of a knowledge claim, and, second, the premiss assertion has the force of a claim to justifiably believe A:

$$\frac{\vdash A}{\vdash \Box^K A} \qquad \frac{\vdash A}{\vdash \Box^B A}$$

With this understanding, the following are also evidently correct:

$$\frac{\vdash \Box^K A \quad \vdash \Box^K [A \supset B]}{\vdash \Box^K B} \qquad \frac{\vdash \Box^B A \quad \vdash \Box^B [A \supset B]}{\vdash \Box^B B}$$

A properly conceived theory of epistemic modal logic is a first-person account enabling a person to understand and explore the consequences of her own knowledge and belief, enlarging her knowledge and belief in the process.

Even ordinary, non-modal, statements made using explicitly first-person expressions call for explicitly first-person theories. Consider Descartes' cogito, either:

> I am thinking, I (now) exist.

or:

> I exist.

Each person's *cogito* is different from that of everyone else; no one can make another person's *cogito*. No one's *cogito* is analytic (its truth conditions *can* fail to be satisfied), and no one's is *a priori* (that my *cogito* is known by me to be true depends on my awareness, my experience, of myself). However, each person is committed, by her self-awareness and her knowledge of the language, to accept her own *cogito*. We need to recognize a new category of statements, and probably a new category of assertive illocutionary acts, to accommodate statements like the *cogito* and illocutionary acts like one's assertion of one's own *cogito*. I propose that we recognize a class of *undeniable* statements. This class would accommodate analytic statements and *a priori* statements as well as essentially first-person statements like the *cogito*. Perhaps the assertion of the *cogito* could be characterized as *analytically compelling*, or *logically compelling*. This class would also include assertions of analytic and *a priori* statements.

Statements made using first-person expressions are essentially and explicitly first-person statements, and require explicitly first-person logical theories to give an adequate account of them. I also recognize an important difference between referring and non-referring uses of singular terms, of both proper names and descriptive singular terms. Most singular terms can be used in both ways, on different occasions, though first person pronouns may have only a referring use. I think that in referring to an object or event, a person exploits a causal connection linking her to that object or event. Everyone's causal connections start with herself, so that no two people can perform the same referring act, and no two people can perform referring acts that are essentially similar to one another. If I am right, an adequate logical treatment of referring will need to be an explicitly first-person logical theory.

As I understand them, all logical theories are first-person theories, but standard theories have an apparently third-person character. For many purposes, it is important to go beyond standard theories, and develop and investigate explicitly first-person theories. Some phenomena cannot be understood and some puzzles cannot be solved without doing this.

References

Kearns, J. T. (2000). An Illocutionary Logical Explanation of the Surprise Execution. *History and Philosophy of Logic, 20,* 195–214.

Kearns, J. T. (2006). Conditional Assertion, Denial, and Supposition as Illocutionary Acts. *Linguistics and Philosophy, 29,* 455–485.

Kearns, J. T. (2007). An Illocutionary Logical Explanation of the Liar Paradox. *History and Philosophy of Logic, 28,* 31–66.

Smith, M. (2003). Neutral and Relative Value after Moore. *Ethics, 113,* 576–598.

Vanderveken, D., & Searle, J. (1985). *Foundations of Illocutionary Logic.* Cambridge: Cambridge University Press.

John T. Kearns
University at Buffalo
USA
E-mail: kearns@buffalo.edu

The Weak Logic of Modal Metaframes

MANFRED KUPFFER[1]

Abstract: While counterpart semantics and ordinary Kripke semantics agree about modality *de dicto*, in counterpart semantics, modality *de re* is treated in a different fashion. One of the main motivations for this was the intuition *de dicto* and *de re*, necessity and essentiality, while being closely related, might have a different modal logic. While modal metaframes provide a version of counterpart semantics for quantified modal logic that is very successful in other respects, the notion of "strong" validity employed by their inventors does not allow to capture this intuition. In this paper I provide a tableaux calculus for a "constant domain" version of metaframes with "weak" validity.

Keywords: quantified modal logic, counterparts, metaframes, weak validity, tableaux

1 Metaframe semantics

There are two main competing approaches to the semantics of quantified modal logic, *viz.* ordinary Kripkean semantics and counterpart semantics.[2] While both agree on the treatment of modality attached to purely general propositions, so-called modality *de dicto* as in **(1)**, they disagree about the treatment of modality *de re* as in **(2)**.

(1) There could be honest politicians.

(2) Berlusconi might return to office.

(1) is uniformly analysed in terms of a relation of accessibility between worlds. Whereas ordinary Kripke semantics applies the same procedure to sentences like **(2)**, counterpart semantics analyses these in terms of a relation of accessibility between possible individuals (in worlds). We will write

---

[1] Thanks to numerous audiences and to Giovanna Corsi, Peter Fritz, Eugenio Orlandelli, and Heinrich Wansing for helpful remarks.

[2] Witness recent textbook treatments of quantified modal logic. Goldblatt (2011) belongs to the Kripkean camp, Gabbay, Shehtman, and Skvortsov (2009) propose a kind of counterpart semantics.

$waRvb$ for "b is, in v, a counterpart for a in w". The relation is employed in a semantics for modality *de re*. According to counterpart-semantics, it holds in w that possibly, Pa, iff there is a world v, and in it, a counterpart b of a in w, such that b has, in v, P.

David Lewis has stressed the importance of the notion of a *way* for the metaphysics of modality. E.g. possible worlds are, or at least represent, maximally specific ways the world might have been in. Likewise counterparts of a in v represent maximally specific ways that a might be in w, see (Kupffer, 2010b). This may serve as a kind of justification of counterpart semantics: clearly, a might be P iff being P is a way a could be in. This in turn is the case exactly iff there is a maximally specific way a might be in, such that every individual that is that way is also P; which in turn holds iff there is a world v and a counterpart of a in v which is P.

While this shows that counterparts are defensible, it does not show that you have to use them. Here, adherents of counterpart semantics point to advantages of their approach. Many of these advantages may be summarised as follows: modality *de re* is special. There are subtle differences between questions of essence and potentiality, i.e. modality *de re*, and questions about how the world must or can be, i.e. modality *de dicto*. Now, if there are such differences, it seems to be a natural thing to treat these questions apart.

E.g. there is *sortal relativity*. Could this table be made of ice? This depends on whether we consider it *qua* table, or *qua* being a table made out of wood.[3] Quite generally, our answers to questions of essence seem to depend on features of the context. This may be taken to explain the flexibility of our judgements, and also the multitude of cases, where we do not have a clue how to answer such questions (Lewis, 1968): in all cases of context-dependence, indeterminacy may occur when context fails to fix the relevant parameters. Neither sortal relativity, widespread indeterminacy, nor a comparable amount of flexibility in our judgements is observed in the case of modality *de dicto*.

There even seem to be *logical differences* between modality *de re* and modality *de dicto*. This has even been taken for granted in the seminal paper about counterpart semantics (Lewis, 1968). Lewis assumes that accessibility is total in the case of modality *de dicto* and not transitive in the case of modality *de re*. As a result, the schema (4), which is seen as a cornerstone of the logic of metaphysical modality *de dicto*, is indeed invalid in the case of modality *de re*. Later he provided an argument in the form of an appeal

---
[3]Cf. (Lewis, 2001).

to Chisholm's so-called paradox:

The rowing boat has been made out of wood. Could it consist entirely of carbon fiber instead? At least sometimes, it seems a natural thing to say that it couldn't.

$$\neg \Diamond Pa$$

Being in fact made out of wood, the boat couldn't consist entirely of carbon fiber, because a boat made entirely out of carbon fiber wouldn't be *this* boat.

On the other hand, even many essentialists about material constitution are ready to admit that the boat could have been made *in part* of carbon fiber. One can always replace a small amount (say, 10 %) of wooden parts by carbon fiber parts without building a different boat altogether. But then

$$\underbrace{\Diamond \ldots \Diamond}_{10 \text{ times}} Pa$$

is true. It is not important that you share these intuitions. It is important that they are coherent. Now, if they are, then the schema (4)

$$\Diamond \Diamond \varphi \to \Diamond \varphi,$$

is *not* logically true, for if it were you could apply it to the formula above (by modus ponens) and after 9 consecutive steps of this kind you would end up with $\Diamond Pa$ in contradiction to the first intuition.[4]

The invalidity of *de re* instances of (4) contrasts with the logical properties of *de dicto* modality, modality attached to purely general propositions. Here, we ask whether something is possible *at all*. And the logic of this very wide kind of possibility (also known as metaphysical possibility) in *de dicto* contexts is uncontroversially **S5** and hence includes the validity of (4) for every completely general φ. If it *could be* possible that pigs fly, then it *is* possible that pigs fly, e.g.

So, *de re* and *de dicto* modal statements are quite different. And yet, both kinds of statement use the same modal operators. We face the following task, then: provide a unified semantics for such modal operators that allows to account for the differences at the same time. Counterpart semantics is an attempt at such a semantics.

A counterpart-relation need not be transitive, even if the relation of accessibility between worlds is transitive. This could enable (4) to hold for

---

[4]Because this failure of (4) is surprising, at least, the above has been termed a *paradox*, cf. (Chisholm, 1967).

modality *de dicto*, but not in the case of modality *de re*. That different counterpart relations are compatible with the same accessibility relation between worlds might also help to account for the remaining differences between the two kinds of modality.

Apart from the distinction between *de re* and *de dicto*, you might also distinguish between modalities in terms of the number of their *res*, see below. In metaframe semantics, see (Gabbay et al., 2009; Skvortsov & Shehtman, 1993), these distinctions lead to different relations of accessibility, as below.

It is possible that
there are honest politicians $\quad wRv$
Berlusconi returns to office $\quad waRvd$
Nečas and Nagyová do not repent $\quad wabRvde$
Anton gives Berta Caesar's regards $\quad wabcRvdef$
etc.

The authors adopt this multitude of accessibility relations for purely technical reasons. But again, counterparts of sequences $\vec{a}$ in w could be taken to represent maximally specific ways, $\vec{a}$ might be in w; and it is a truism that $\vec{a}$ stands possibly in relation P iff every maximally specific way $\vec{a}$ might be in, is, among other things a way to be P-related. So the generalised counterpart relation in metaframe semantics is justifiable in precisely the same way as the ordinary counterpart relation between individuals. On the other hand, ordinary counterpart semantics uses simple counterparts only, and one might ask at this point, whether we need counterparts of sequences at all.

Indeed counterpart of pairs determine counterpart relations between their component parts. But the reverse need not be the case. E.g. a pair of counterparts (a', b') of a and b might lack certain relations which are essential to the pair (a, b). Take e.g. the father-son relation. Hazen (1979) asks us to consider a world with two different counterparts of a father-son-pair. While every father of such a pair is a counterpart of the actual father, and every son of such a pair is a counterpart of the actual son, there are mixed pairs of these counterparts that do not constitute a father-son pair, and are hence no counterparts of the actual father-son pair (supposing the father-son relation is essential). For this reason, Hazen argues, counterparts of sequences

Weak Logic of Modal Metaframes

are needed in addition to counterparts of individuals. Let us now define this multiplicity of relations of accessibility in a formal way. We adopt the following conventions: $I^0 := \emptyset$, $I^n := \{1,\ldots,n\}$. An n-place *sequence* over B is a function from I^n into B ($\emptyset$ is the 0-place sequence). B^n is the set of n-place sequences over B. Suppose $\vec{a} \in B^n$. σ is an m-place *transformation* of $\vec{a}$ iff σ is a function from I^m into I^n. Then, $\vec{a}\sigma = \vec{a}\cdot\sigma = (a_{\sigma(1)},\ldots,a_{\sigma(m)})$. If $\vec{a}$ is an n-place and $\vec{b}$ an m-place sequence, then $\vec{a}\sqcup\vec{b}$ is the $n+m$-place sequence obtained by appending $\vec{b}$ to $\vec{a}$. We sometimes use a also to denote $\{(1,a)\}$, the corresponding one-place sequence.

Definition 1 *A (constant domain) metaframe is a triple $(W, D, (R_n)_{n\in\mathbb{N}})$, s.t. W and D are non-empty sets and for every $n \in \mathbb{N}$ $R_n \subseteq (W \times D^n)^2$.*

(Multiple) de re modality is subject to certain closure properties. E.g. if it is possible that a sees b, it is also possible that b is seen by a, furthermore that a sees someone, and that b is seen by someone. Finally, in the same kind of situation it is also possible that a sees someone and c is self-identical. To account for such dependencies we assume that metaframes are *modal*.

Definition 2 *A metaframe $(W, D, (R_n)_{n\in\mathbb{N}})$ is modal iff C.1 and C.2 below hold for arbitrary $\vec{a}, \vec{c} \in D^n$, $\vec{b} \in D^m$, and m-place transformations σ of $\vec{a}$.*

C.1 *If $w\vec{a}R_n v\vec{c}$, then $w\vec{a}\sigma R_m v\vec{c}\sigma$.*

C.2 *If $w\vec{a}\sigma R_m v\vec{b}$, then there is a $\vec{c}$, s.t. $\vec{b} = \vec{c}\sigma$ and $w\vec{a}R_n v\vec{c}$.*

We will now interpret the language of quantified modal logic in models based on modal metaframes. For technical reasons we use a version of the language without open formulae and with no quantifier Qv occurring in the scope of a quantifier $Q'v$ with the same variable v. The language includes atomic formulas (0-place predicates).

Definition 3 *$\mathcal{M}$ is a model if, and only if $\mathcal{M} = (W, D, (R_n)_{n\in\mathbb{N}}, V)$, where $(W, D, (R_n)_{n\in\mathbb{N}})$ is a modal metaframe and V a function, s.t. if P is an n-place predicate, then $V(P) \subseteq W \times D^n$ and if α is a name, then $V(\alpha) \in D$.*

If α is a name and $\mathcal{M} = (W, D, (R_n)_{n\in\mathbb{N}}, V)$ is a model, then $\alpha^\mathcal{M} := V(\alpha)$ and if $\vec{\alpha}$ is an n-place sequence of names, then $\vec{\alpha}^\mathcal{M} \in D^n$ is the sequence of their values $\alpha_i^\mathcal{M}$.

If $\mathcal{M} = (W, D, (R_n)_{n \in \mathbb{N}}, V)$ is a model, α an n-place sequence of names and $\vec{b} \in D^n$, then $\mathcal{M}[\vec{b}/\vec{\alpha}] := (W, D, (R_n)_{n \in \mathbb{N}}, V')$, where V' is exactly like V, except that $V'(\alpha_i) = b_i$ for all i ($1 \leq i \leq n$).

In the following $\mathcal{M} = (W, D, (R_n)_{n \in \mathbb{N}}, V)$ is an arbitrary metaframe model, $w \in W$, P is an n-place predicate, and $\alpha_1, \ldots, \alpha_n$ are names. Let φ and ψ be arbitrary formulae. Let α be a name, and let v be a variable which does not appear in φ. Truth in a model and a world is defined as follows.

$\mathcal{M} \models P\alpha_1 \ldots \alpha_n : w$ iff $(w, \alpha_1^{\mathcal{M}}, \ldots, \alpha_n^{\mathcal{M}}) \in V(P)$,
$\mathcal{M} \models \neg\varphi : w$ iff $\mathcal{M} \not\models \varphi : w$,
$\mathcal{M} \models (\varphi \wedge \psi) : w$ iff $\mathcal{M} \models \varphi : w$ and $\mathcal{M} \models \psi : w$,
$\mathcal{M} \models (\varphi \vee \psi) : w$ iff $\mathcal{M} \models \varphi : w$ or $\mathcal{M} \models \psi : w$,
$\mathcal{M} \models (\varphi \to \psi) : w$ iff $\mathcal{M} \not\models \varphi : w$ or $\mathcal{M} \models \psi : w$,
$\mathcal{M} \models \forall v \varphi[v/\alpha] : w$ iff   for all $a \in D$: $\mathcal{M}[a/\alpha] \models \varphi : w$,
$\mathcal{M} \models \exists v \varphi[v/\alpha] : w$ iff   for some $a \in D$: $\mathcal{M}[a/\alpha] \models \varphi : w$,
$\mathcal{M} \models \Box\varphi : w$ iff   for all $v \in W$ and every $\vec{b} \in D^n$:
     if $w\vec{\alpha}^{\mathcal{M}} R v \vec{b}$, then $\mathcal{M}[\vec{b}/\vec{\alpha}] \models \varphi : v$,
$\mathcal{M} \models \Diamond\varphi : w$ iff   there is a $v \in W$ and a $\vec{b} \in D^n$, s.t.
     $w\vec{\alpha}^{\mathcal{M}} R v \vec{b}$ and $\mathcal{M}[\vec{b}/\vec{\alpha}] \models \varphi : v$.

In the last two clauses, $\vec{\alpha}$ is the n-place sequence of names from φ in the order of their first occurrence in φ.

2 Logic

We will define validity as usual in modal logic: a formula φ is valid in a model $\mathcal{M} = (W, D, (R_n)_{n \in \mathbb{N}}, V)$ iff $\mathcal{M} \models \varphi : w$ for all $w \in W$. φ is valid in a class of $\mathcal{F}$ metaframes iff for all models $\mathcal{M}$ over some frame $F \in \mathcal{F}$: φ is valid in $\mathcal{M}$.

Lemma 1 *For all models $\mathcal{M} = (W, D, (R_n)_{n \in \mathbb{N}}, V)$ and all n-place formulas φ, $w \in W$ and n-place sequences of names $\vec{\alpha}$ and $\vec{\beta}$:*

$$\mathcal{M}[\vec{\beta}^{\mathcal{M}}/\vec{\alpha}] \models \varphi : w \text{ iff } \mathcal{M} \models \varphi[\vec{\beta}/\vec{\alpha}] : w.$$

Corollary 1 *If φ is a formula, $\vec{\gamma}$ an n-place sequence of names not occurring in φ and $\vec{a} \in D^n$, then $\mathcal{M}[\vec{a}/\vec{\gamma}] : w \models \varphi$ iff $\mathcal{M} \models \varphi : w$.*

A metaframe $(W, D, (R_n)_{n \in \mathbb{N}})$ is *0-transitive* iff R_0 is transitive, and *1-transitive* iff R_1 is. It holds that 1-transitivity implies 0-transitivity but not vice versa. Let $\mathcal{F}$ be the class of 0-transitive frames. Now consider:

4_0 $\Box p \to \Box\Box p$,

4_1 $\Box Pa \to \Box\Box Pa$.

4_0 is valid in $\mathcal{F}$, 4_1 not. This means that uniform substitution of a formula for an atomic formula does not always preserve validity in $\mathcal{F}$. Since we treat propositional variables as 0-place predicates, the same example also shows that second-order substitution (substitution of an (open) formula for a predicate) sometimes breaks down in metaframe semantics.

For the inventors of these semantics, these were unwelcome results. The reason is that the technical literature on QML is often concerned with the question of finding semantics for a given logic, and the definition of their axiom systems includes substitution. Hence they proposed to combine metaframe semantics with *strong validity*, where a formula is strongly valid iff every 2nd order substitution instance is valid. E.g. neither 4_1 nor 4_0 are strongly valid in the class of 0-transitive frames.

While this maneuver does indeed save uniform substitution, it looks rather *ad hoc*.[5] Logical truth and logical consequence are semantic notions; *if* logic possesses a certain structural feature it should possess a semantic explanation. Simply importing structural notions into semantics cannot provide such an explanation.

Additionally, in the present case there are reasons to believe that our logic should *not* possess the structural feature under consideration. Metaframe semantics treats the distinction between *de re* and *de dicto* as a syntactical one, in terms of whether names or open variables appear in the scope of a modal operator.[6] This means that 4_0 is *de dicto*, whereas 4_1 is *de re*. Above I have argued that *de dicto* and *de re* should be permitted to have a different logic; hence an adequate notion of validity should allow for substitution failures rather than rule them out.

This is why we stick to ordinary "weak" validity. But then, we now need a logical calculus that does not have substitution built in. In the following I will present a suitable tableaux calculus.

---

[5]Cf. (Bauer & Wansing, 2002).
[6]This feature is shared by Lewis' seminal paper. For a critique, and a counterpart semantics without this feature, see my (Kupffer, 2010a).

The rules for first order logic are standard:

$$\frac{\neg\neg\varphi : \omega}{\varphi : \omega}$$

$$\frac{\varphi \wedge \psi : \omega}{\varphi : \omega \\ \psi : \omega} \qquad \frac{\neg(\varphi \wedge \psi) : \omega}{\neg\varphi : \omega \mid \neg\psi : \omega}$$

$$\frac{\varphi \vee \psi : \omega}{\varphi : \omega \mid \psi : \omega} \qquad \frac{\neg(\varphi \vee \psi) : \omega}{\neg\varphi : \omega \\ \neg\psi : \omega}$$

$$\frac{\varphi \rightarrow \psi : \omega}{\neg\varphi : \omega \mid \psi : \omega} \qquad \frac{\neg(\varphi \rightarrow \psi) : \omega}{\varphi : \omega \\ \neg\psi : \omega}$$

For the following rules, α is a name, β is a name on the same branch, or $\beta = a_1$, if no name occured on the branch so far; γ is a *new* name, i.e. a name, which did not occur on the branch so far.

$$\frac{\forall v\, \varphi[v/\alpha] : \omega}{\varphi[\beta/\alpha] : \omega} \qquad \frac{\neg\forall v\, \varphi[v/\alpha] : \omega}{\neg\varphi[\gamma/\alpha] : \omega}$$

$$\frac{\exists v\, \varphi[v/\alpha] : \omega}{\varphi[\gamma/\alpha] : \omega} \qquad \frac{\neg\exists v\, \varphi[v/\alpha] : \omega}{\neg\varphi[\beta/\alpha] : \omega}$$

We will now provide rules for the modal operators. In the following $\vec{\alpha}$, $\vec{\beta}$ and $\vec{\gamma}$ are n-place sequences of pairwise distinct names, $\vec{\gamma}$ is a sequence of *new* names, σ is a transformation of an n-place sequence, ω and η are arbitrary world-parameters, ζ a new one. If φ is a formula and $\vec{\alpha}$ is the sequence of names in φ in the order of their first appearance, instead of φ we also write $\varphi[\vec{\alpha}]$.

Weak Logic of Modal Metaframes

$$\frac{\Box\psi[\vec{\alpha}\sigma] : \omega \quad w\vec{\alpha}R\eta\vec{\beta}}{\psi[\vec{\beta}/\vec{\alpha}] : \eta}$$

$$\frac{\neg\Box\varphi[\vec{\alpha}] : \omega}{w\vec{\alpha}R\zeta\vec{\gamma} \quad \neg\varphi[\vec{\gamma}/\vec{\alpha}] : \zeta}$$

$$\frac{\Diamond\varphi[\vec{\alpha}] : \omega}{w\vec{\alpha}R\zeta\vec{\gamma} \quad \varphi[\vec{\gamma}/\vec{\alpha}] : \zeta}$$

$$\frac{\neg\Diamond\psi[\vec{\alpha}\sigma] : \omega \quad w\vec{\alpha}R\eta\vec{\beta}}{\neg\psi[\vec{\beta}/\vec{\alpha}] : \eta}$$

One final rule completes the basic calculus **QK**. Let α be a name occurring on the branch, but not in $\vec{\alpha}$. Let γ be a new name.

r-Ext

$$\frac{w\vec{\alpha}R\eta\vec{\beta}}{w\vec{\alpha}\sqcup\alpha R\eta\vec{\beta}\sqcup\gamma}$$

(This rule should only be applied, if there are no names γ, s.t. $w\vec{\alpha}\sqcup\alpha R\eta\vec{\beta}\sqcup\gamma$ is already on the branch.)

Various additional rules might be added, resulting in different extensions of **QK**. Here are some generalisations of standard examples.

In the following, $\vec{\alpha}$, $\vec{\beta}$ and $\vec{\gamma}$ are n-place sequences of names. The first rule should only be applied if $\vec{\alpha}$ has already occured on the branch above.

n-Refl

$$\frac{\varphi : \omega}{w\vec{\alpha}R_n w\vec{\alpha}}$$

n-Sym

$$\frac{w\vec{\alpha}R_n\eta\vec{\beta}}{\eta\vec{\beta}R_n w\vec{\alpha}}$$

n-Trans

$$\frac{w\vec{\alpha}R_n\eta\vec{\beta} \quad \eta\vec{\beta}R_n\zeta\vec{\gamma}}{w\vec{\alpha}R_n\zeta\vec{\gamma}}$$

Derivability is defined as usual: a branch is *closed* iff it contains a formula and its negation. A tree is *closed* iff all of its branches are. A formula φ is *derivable* in a calculus X, iff there is a closed tree of X with $\neg\varphi : w$ as its root. Derivability in **QK** is sound and complete in the class of all metaframes, see the appendix.

Appendix

Soundness of QK

Let B be a branch, let W^B be the set of its world-parameters and let $\mathcal{M} = (W, D, (R_n)_{n \in \mathbb{N}}, V)$ be a model. Call f an *interpretation* of W^B in $\mathcal{M}$ iff $f \colon W^B \to W$. Let f be an interpretation. Say that f *satisfies* $\varphi : w$ in $\mathcal{M}$ iff $\mathcal{M} \models \varphi : f(w)$, f satisfies $w\vec{\alpha}R_n v\vec{\beta}$ in $\mathcal{M}$ iff $f(w)\vec{\alpha}^{\mathcal{M}} R_n f(v)\vec{\beta}^{\mathcal{M}}$, f *satisfies* B in $\mathcal{M}$ iff f satisfies all nodes on B in $\mathcal{M}$. B is *satisfiable* in the class of frames $\mathcal{F}$, iff there is a model $\mathcal{M}$ over some frame $F \in \mathcal{F}$ and an interpretation f, s.t. f satisfies B in $\mathcal{M}$; and B is *satisfiable* iff B satisfiable in the class of all metaframes.

Theorem 1 *If a branch B is satisfiable, and some rule is applied to B, then at least one of the resulting branches is satisfiable.*

Proof. By an inspection of the tableaux rules. □

Corollary 2 (Soundness) φ *is* satisfiable $\Rightarrow \varphi$ *is* **QK***-consistent.*

Further correctness results are easily obtainable.

Example 1 φ is *satisfiable* in the class of 0-transitive frames $\Rightarrow \varphi$ is **QK+0-Trans**-consistent.

Completeness of QK

Lemma 2 *For every formula φ and every world-parameter ω there is a complete tree of X, which begins with $\varphi : \omega$.*

Proof. by outlining a systematic procedure, which secures for every branch B of a tree that all rules of the calculus X that could have been applied to B have in fact been applied to B. □

Definition 4 *Let B be a complete open branch. $\mathcal{M}^B$, the model induced by B, is defined to be $(W^B, D^B, (R^B)_{n \in \mathbb{N}}, V^B)$, where W^B is the set of world-parameters on B and D^B the set the names on B; if no name occurs on the branch $D^B = \{a_1\}$. If ω and η are world-parameters and $\vec{\alpha}$ and $\vec{\beta}$ n-place sequences of names, then $\omega\vec{\alpha}R_n^B\eta\vec{\beta}$ iff there are $\vec{\gamma}, \vec{\delta} \in D^{B^m}$ and an n-place transformation σ of $\vec{\gamma}$, s.t. $\vec{\alpha} = \vec{\gamma}\sigma$ and $\vec{\beta} = \vec{\delta}\sigma$ and $\omega\vec{\gamma}R\eta\vec{\delta}$ occurs on B. If ω and $\vec{\alpha}$ are as above and φ is an n-place predicate, then it holds: $\vec{\alpha} \in V_\omega^B(\varphi)$ iff $\varphi\vec{\alpha} : \omega$ occurs on B.*

Weak Logic of Modal Metaframes

Lemma 3 *Let B a complete open branch and R^B be as defined above. If $w\vec{\alpha}R^B\eta\vec{\beta}$ and σ is a transformation of $\vec{\alpha}$, then $w\vec{\alpha}\sigma R^B\eta\vec{\beta}\sigma$.*

Proof. Suppose $w\vec{\alpha}R^B\eta\vec{\beta}$. Let σ be a transformation of $\vec{\alpha}$. Then there is a $\vec{\gamma}$ and a $\vec{\delta}$, as well as a transformation τ, s.t. $w\vec{\gamma}R\eta\vec{\delta}$ on B and $\vec{\alpha} = \vec{\gamma}\tau$ and $\vec{\beta} = \vec{\delta}\tau$. Therefore, $w\vec{\gamma}\tau\sigma R^B\eta\vec{\delta}\tau\sigma$ (because $\tau\sigma = \tau \cdot \sigma$ is a transformation) and hence $w\vec{\alpha}\sigma R^B\eta\vec{\beta}\sigma$. □

Let $F = (W, D, (R_n)_{n\in\mathbb{N}})$ be a metaframe. R_n is *functional* iff it holds that if $w\vec{a}R_n v\vec{b}$ and $a_i = a_j$, then $b_i = b_j$. F is functional iff all R_n are. F *allows for extensions* iff it holds that if $w\vec{a}R_n v\vec{b}$ and $a \in D$, then there is a $b \in D$, s.t. $w\vec{a}_{\sqcup}aR_{n+1}v\vec{b}_{\sqcup}b$.

Lemma 4 *A metaframe $\mathcal{F}$ is modal iff it is functional, allows for extensions, and C.1 holds.*

Lemma 5 *R^B is functional.*

Proof. Suppppose $w\vec{\alpha}R^B\eta\vec{\beta}$. Then there are $\vec{\gamma}, \vec{\delta}$ and a transformation σ, s.t. $\vec{\alpha} = \vec{\gamma}\sigma$ and $\vec{\beta} = \vec{\delta}\sigma$ and $w\vec{\gamma}R\eta\vec{\delta}$ occurs on B. By the tableaux rules: if $w\vec{\gamma}R\eta\vec{\delta}$ is on B, then $\vec{\gamma}$ and $\vec{\delta}$ are injective sequences. Therefore

$$\alpha_i(=\gamma_{\sigma i}) = \alpha_j(=\gamma_{\sigma j}) \text{ iff } \beta_i(=\delta_{\sigma i}) = \beta_j(=\delta_{\sigma j}).$$

□

Lemma 6 *If $w\vec{\alpha}R^B\eta\vec{\gamma}$ and $\alpha \in D^B$ does not occur in $\vec{\alpha}$, then there is a γ that does not occur in $\vec{\gamma}$, s.t. $w\vec{\alpha} + \alpha R^B\eta\vec{\gamma} + \gamma$.*

Proof. By Lemma 3 and r-Ext □

Lemma 7 *$\mathcal{M}^B$ is a model over a modal metaframe.*

Proof. It follows from the above and Lemma 4 that the metaframe of $\mathcal{M}^B$ is modal. Obviously, $\mathcal{M}^B$ is a model. □

Lemma 8 *Let B be a complete open branch. (i) for all ω: if $\varphi : \omega$ occurs on B, then $\mathcal{M}^B \models \varphi : \omega$; (ii) for all ω: if $\neg\varphi : \omega$ occurs on B, then $\mathcal{M}^B \not\models \varphi : \omega$.*

Proof. By induction on φ, e.g.

(i) $\Box$: Let $\Box\varphi[\vec{\alpha}] : \omega$ be on B. We want to show that for all $\vec{\beta} \in D^B$ and $\eta \in W^B$, if $\omega\vec{\alpha}R^B\eta\vec{\beta}$, then $\mathcal{M}^B[\vec{\beta}/\vec{\alpha}] \models \varphi : \eta$. Suppose $\omega\vec{\alpha}R^B\eta\vec{\beta}$. Because of Def. R^B there are $\vec{\alpha}'$, $\vec{\beta}'$ and a transformation σ of $\vec{\alpha}'$, s.t. $\omega\vec{\alpha}'R\eta\vec{\beta}'$ on B, $\vec{\alpha} = \vec{\alpha}'\sigma$ as well as $\vec{\beta} = \vec{\beta}'\sigma$. Because B is closed, then $\varphi[\vec{\beta}'/\vec{\alpha}'] : \eta \ (= \varphi[\vec{\beta}/\vec{\alpha}] : \eta)$ is on B, and by induction hypothesis it follows that $\mathcal{M}^B \models \varphi[\vec{\beta}/\vec{\alpha}] : \eta$. With Lemma 1 it follows that $\mathcal{M}^B[\vec{\beta}/\vec{\alpha}] \models \varphi : \eta$.

(ii) $\Box$: Let $\neg\Box\varphi[\vec{\alpha}] : \omega$ on B. Then there is a $\vec{\gamma} \in D^B$ and $\zeta \in W^B$, s.t. $\omega\vec{\alpha}R\zeta\vec{\gamma}$ and $\neg\varphi[\vec{\gamma}/\vec{\alpha}] : \zeta$ on B. Therefore, $\omega\vec{\alpha}R^B\zeta\vec{\gamma}$ (because of Def. R^B); and $\mathcal{M}^B \not\models \varphi[\vec{\gamma}/\vec{\alpha}] : \zeta$ (by induction hypothesis). With Lemma 1 it follows that $\mathcal{M}^B[\vec{\gamma}/\vec{\alpha}] \not\models \varphi : \zeta$, hence $\mathcal{M}^B \not\models \Box\varphi : \omega$. $\square$

Theorem 2 (Completeness) φ *is* consistent $\Rightarrow \varphi$ *is satisfiable.*

Proof. Suppose φ is *consistent*. Then, in each tree, beginning with $\varphi : \omega$ there is an open branch. Additionally there is a complete branch, beginning with $\varphi : \omega$. Consider an open branch B in a *complete* tree beginning with $\varphi : \omega$. Because φ is on B, it holds with Lemma 8 that $\mathcal{M}^B \models \varphi : \omega$. Therefore φ is satisfiable. $\square$

Further completeness results are easily obtainable:

Example 2 If φ is **QK+0-Trans**-consistent, then it is satisfiable in the class of 0-transitive metaframes.

In order to prove this we have to show that the induced model has the relevant frame-property, e.g. the induced model of any complete open branch of a **QK+0-Trans**-tree is 0-transitive.

References

Bauer, S., & Wansing, H. (2002). Consequence, Counterparts and Substitution. *The Monist*, *85*, 483–497.

Chisholm, R. M. (1967). Identity through Possible Worlds: Some Questions. *Noûs*, *1*, 1–8.

Gabbay, D., Shehtman, V., & Skvortsov, D. P. (2009). *Quantification in Nonclassical Logic*. New York: Elsevier.

Goldblatt, R. (2011). *Quantifiers, Propositions and Identity*. Cambridge: Cambridge University Press.

Hazen, A. P. (1979). Counterpart-theoretic Semantics for Modal Logic. *Journal of Philosophy, 76*, 319–338.

Kupffer, M. (2010a). *Counterparts and Qualities.* PhD Thesis, University of Konstanz. Retrieved from `http://nbn-resolving.de/urn:nbn:de:bsz:352-opus-107638`

Kupffer, M. (2010b). Counterpart Semantics and the Multiple De Re. In G. Imaguire & D. Jacquette (Eds.), *Possible Worlds.* Munich: Philosophia.

Lewis, D. K. (1968). Counterpart Theory and Quantified Modal Logic. *Journal of Philosophy, 65*, 113–126.

Lewis, D. K. (2001). Truthmaking and Difference-Making. *Noûs, 35*, 602–615.

Skvortsov, D. P., & Shehtman, V. B. (1993). Maximal Kripke-type Semantics for Modal and Superintuitionisic Predicate Logics. *Annals of Pure and Applied Logic, 63*, 69–101.

Manfred Kupffer
Goethe-University Frankfurt
Germany
E-mail: `kupffer@em.uni-frankfurt.de`

Compliance and Pure Erotetic Implication

PAWEŁ ŁUPKOWSKI[1]

Abstract: Notions of compliance and pure erotetic implication are introduced and compared in the context of grasping the dependency between questions.

Keywords: question dependency, erotetic implication, compliance, IEL, inquisitive semantics

1 Introduction

In this paper I will be addressing a situation where a question is given as a response to a question. I will present how this type of reply to a question—which I will call a *question response*—is grasped with the notion of compliance (developed within inquisitive semantics—INQ, cf. (Groenendijk, 2009; Groenendijk & Roelofsen, 2011)) and by the notion of erotetic implication (which comes from Inferential Erotetic Logic—IEL framework, cf. (Wiśniewski, 1995, 2013)).

Łupkowski and Ginzburg (2013) describe a typology of question responses based on the British National Corpus (BNC) study. Among these various question responses one type is especially interesting, namely *dependent question* responses. The rationale behind dependent questions can be summarised as follows (Ginzburg, 2012, p. 57): if Q depends on Q_1 then discussions of Q_1 will necessarily bring about the provision of information about Q.

Definition 1 (Ginzburg, 2012, p. 57) Q depends on Q_1 *iff any proposition p such that p resolves Q_1, also satisfies p entails r such that r is about Q. This allows to say that Q_1 can be used to respond to Q if Q depends on Q_1.*

---

[1] I would like to give my thanks to M. Urbański for helpful feedback and comments on a draft of this article. This work was supported by funds of the National Science Council, Poland (DEC 2012/04/A/HS1/00715).

In other words, Q_1 is an acceptable response to Q. One may easily notice that this type of question response captures the intuition of introducing a sub-question which will somehow help to retrieve the answer to the initial question. The following example illustrate this idea:

> A: Do you want me to <*pause*> push it round?
> B: Is it really disturbing you? [FM1, 679–680][2]
> (i.e. *Whether I want you to push it depends on whether it really disturbs you*)

Let us now take a closer look at how this relation between questions is expressed in INQ and IEL.

2 The notion of compliance

In the framework of inquisitive semantics, the dependency relation is analysed in terms of *compliance*. The intuition behind the notion of compliance is to provide a criterion to 'judge whether a certain conversational move makes a significant contribution to resolving a given issue' (Groenendijk & Roelofsen, 2011, p. 167). If we take two conversational moves: the initiative A and the response B, there are two ways in which B may be compliant with A (cf. Groenendijk & Roelofsen, 2011, p. 168):

(a) B may partially *resolve* the issue raised by A (answerhood).

(b) B may replace the issue raised by A by an easier to answer sub-issue (subquestionhood).[3]

Here I will be interested only in the case when we are dealing with subquestionhood. Before I will be able to give the definition of compliance, first I will introduce (after Wiśniewski and Leszczyńska-Jasion (2013, pp. 6–12)) the necessary concepts of INQ, especially the notion of question used in this framework.

---

[2]This notation indicates the British National Corpus file (FM1) together with the sentence numbers (679–680).

[3]In the inquisitive semantics also combinations of (a) and (b) are possible, i.e. B may partially resolve the issue raised by A and replace the remaining issue by an easier to answer sub-issue.

2.1 INQ basic concepts

Firstly, let us introduce a language $\mathcal{L}_\mathcal{P}$. It is a propositional language over a non-empty set of propositional variables $\mathcal{P}$, where $\mathcal{P}$ is either finite or countably infinite. The primitive logical constants of the language are: $\bot, \vee, \wedge, \rightarrow$. Well-formed formulas (wffs) of $\mathcal{L}_\mathcal{P}$ are defined as usual.

The letters A, B, C, D, are metalanguage variables for wffs of $\mathcal{L}_\mathcal{P}$, and the letters X, Y are metalanguage variables for sets of wffs of the language. The letter p is used below as a metalanguage variable for propositional variables.

$\mathcal{L}_\mathcal{P}$ is associated with the set of suitable possible worlds, $\mathcal{W}_\mathcal{P}$, being the *model* of $\mathcal{L}_\mathcal{P}$. A possible world is identified with indices (that is valuations of $\mathcal{P}$). $\mathcal{W}_\mathcal{P}$ is the set of all indices.

A *state* is a subset of $\mathcal{W}_\mathcal{P}$ (states are thus sets of possible worlds). I will use the letters σ, τ, γ, to refer to states.

The most important semantic relation between states and wffs is that of *support*. In the case of INQ support, $\succ$, is defined by:

Definition 2 (Wiśniewski & Leszczyńska-Jasion, 2013, p. 6)
Let $\sigma \subseteq \mathcal{W}_\mathcal{P}$.

1. $\sigma \succ p$ *iff for each* $w \in \sigma$: *p is true in* w,[4]
2. $\sigma \succ \bot$ *iff* $\sigma = \emptyset$,
3. $\sigma \succ (A \wedge B)$ *iff* $\sigma \succ A$ *and* $\sigma \succ B$,
4. $\sigma \succ (A \vee B)$ *iff* $\sigma \succ A$ *or* $\sigma \succ B$,
5. $\sigma \succ (A \rightarrow B)$ *iff for each* $\tau \subseteq \sigma$: *if* $\tau \succ A$ *then* $\tau \succ B$.

For our analysis we will use also the notion of the *truth set* of a wff A (in symbols: $|A|$). It is the set of all the worlds from $\mathcal{W}_\mathcal{P}$ in which A is true, where the concept of truth is understood classically.

Now we can introduce the concept of a *possibility* for a wff A. Intuitively it is a maximal state supporting A. This might be expressed as follows:

Definition 3 (Wiśniewski & Leszczyńska-Jasion, 2013, p. 9)
A possibility for wff A *is a state* $\sigma \subseteq \mathcal{W}_\mathcal{P}$ *such that* $\sigma \succ A$ *and for each* $w \notin \sigma : \sigma \cup \{w\} \not\succ A$.

---

[4]"p is true in w" means "the value of p under w equals **1**".

I will use $\lfloor A \rfloor$ to refer to the set of all possibilities for a wff A.

In INQ we may divide all wffs into assertions and inquisitive wffs. The latter are the most interesting from our point of view, because they raise an issue to be solved. When a wff is inquisitive, the set of possibilities for that formula comprises at least two elements. (When a formula has only one possibility it is called assertion.)

Let us now consider a simple example of an inquisitive formula:

$$(p \vee q) \vee \neg(p \vee q) \tag{1}$$

The set of possibilities for (1) is:

$$\{|p|, |q|, |\neg p| \cap |\neg q|\} \tag{2}$$

and its union is just $\mathcal{W}_\mathcal{P}$.

Observe that the language $\mathcal{L}_\mathcal{P}$ does not include a separate syntactic category of questions. However, some wffs are regarded as *having the property of being a question*, or $\mathcal{Q}$-property for short.

Definition 4 (Wiśniewski & Leszczyńska-Jasion, 2013, p. 11)
A wff A of $\mathcal{L}_\mathcal{P}$ has the $\mathcal{Q}$-property iff $|A| = \mathcal{W}_\mathcal{P}$.

Where $\mathcal{W}_\mathcal{P}$ stands for the model of $\mathcal{L}_\mathcal{P}$, and $|A|$ for the truth set of wff A in $\mathcal{W}_\mathcal{P}$. An example of a formula having the $\mathcal{Q}$-property is the formula (1). Hence a wff A is (i.e. has the property of being) a question just in case when A is true in each possible world of $\mathcal{W}_\mathcal{P}$, the wffs having the $\mathcal{Q}$-property are just classical tautologies.

2.2 Compliance

Let Q be an *initiative* and Q_1 a *response* to the initiative. We also assume that Q and Q_1 are inquisitive formulae and that they have the $\mathcal{Q}$-property (further on I will just call them questions for simplicity). $\lfloor Q \rfloor$ denotes the set of possibilities for Q.

Definition 5 (cf. Groenendijk, 2009, p. 22) *Let $\lfloor Q \rfloor = \{|A_1|, \ldots, |A_n|\}$ and $\lfloor Q_1 \rfloor = \{|B_1|, \ldots, |B_m|\}$. Q_1 is compliant with Q (written in symbols $Q_1 \propto Q$), iff*

1. *For each $|B_i|$ ($1 \leq i \leq m$) there exist $k_1, \ldots, k_l$ ($1 \leq k_p \leq n$; $1 \leq p \leq l$) such that $|A_{k_1}| \cup \ldots \cup |A_{k_l}| = \bigcup_{p=1}^{l} |A_{k_p}| = |B_i|$.*

Compliance and Pure Erotetic Implication

2. For each $|A_j|$ $(1 \leq j \leq n)$ there exists $|B_k|$ $(1 \leq k \leq m)$, such that $|A_j| \subset |B_k|$.

As it may be observed—in the case of compliance—we cannot say anything about declarative premises involved in going from a question to question response. The relationship captured by the compliance is a pure sub-questionhood relationship. A simple example of question–question response where the reply is compliant to the initiative illustrates this idea.

Example 1 Q is 'Is John coming to the party and can I come?' while Q_1 is 'Is John coming to the party'. Q_1 may be expressed in INQ as $p \vee \neg p$ and Q as $(p \vee \neg p) \wedge (q \vee \neg q)$. $\lfloor Q \rfloor = \{|p \wedge q|, |\neg p \wedge q|, |p \wedge \neg q|, |\neg p \wedge \neg q|\}$ and $\lfloor Q_1 \rfloor = \{|p|, |\neg p|\}$. It is the case that $Q_1 \propto Q$, because both conditions for compliance are met. For the first condition observe that $|p| = |p \wedge q| \cup |p \wedge \neg q|$ and $|\neg p| = |\neg p \wedge q| \cup |\neg p \wedge \neg q|$. For the second condition let us observe that: $|p \wedge q| \subset |p|$; $|p \wedge \neg q| \subset |p|$; $|\neg p \wedge q| \subset |\neg p|$; $|\neg p \wedge \neg q| \subset |\neg p|$.

3 Erotetic implication

3.1 Language $\mathcal{L}_?$

In the following I will use the formal language $\mathcal{L}_?$. $\mathcal{L}_?$ is First-order Logic language enriched with the question-forming operator ? and brackets {, }. Well formed formulae of FoL (defined as usual) are *declarative well-formed formulae* of $L_?$ (d-wffs for short). Expressions of the form $?\{A_1, \ldots, A_n\}$ are *questions* of $L_?$ (e-formulae) provided that $A_1, \ldots, A_n$ are syntactically distinct d-wffs and that $n > 1$. The set $\mathbf{d}Q = \{A_1, \ldots, A_n\}$ is the set of all *direct answers* to the question $Q = ?\{A_1, \ldots, A_n\}$. The question $?\{A_1, \ldots, A_n\}$ might be read as 'Is it the case that A_1 or is it the case that $A_2, \ldots,$ or is it the case that A_n?'.

For brevity, I will adopt a different notation for one type of questions. So called *(binary) conjunctive questions*,[5] namely $?\{A \wedge B, A \wedge \neg B, \neg A \wedge B, \neg A \wedge \neg B\}$ will be written as $? \pm |A, B|$ ('Is it the case that A and is it the case that B?')—cf. (Wiśniewski, 2003, p. 399).

3.2 Pure erotetic implication

A definition of the erotetic implication (e-implication) can be formulated as follows:

---

[5] For a generalised definition of conjunctive questions see (Urbański, 2001, p. 76).

Definition 6 *A question Q_1 is e-implied by a question Q on the basis of a set X of declarative formulae ($Q_1 \triangleright_X Q$) iff:*

1. *for each direct answer A to the question Q: $X \cup \{A\}$ entails the disjunction of all the direct answers to the question Q_1, and*

2. *for each direct answer B to the question Q_1 there exists a non-empty proper subset Y of the set of direct answers to the question Q such that $X \cup \{B\}$ entails the disjunction of all the elements of Y.*

Intuitively, erotetic implication ensures the following: (i) if Q is sound[6] and X consists of truths, then Q_1 has a true direct answer as well ('transmission of soundness and truth into soundness'—Wiśniewski, 2003, p. 401), and (ii) each direct answer to Q_1, if true, and if all elements of X are true, it narrows down the class in which a true direct answer to Q can be found ('open-minded cognitive usefulness'—Wiśniewski, 2003, p. 402).

It can be observed that e-implication enables the capturing of the relationship between question-question response on the basis of a set of declarative premises. This is not the case for the compliance (see Definition 5)—the reason for this is that there we are only interested in pure questioning without introducing any information. If X is empty, an e-implication ($Q_1 \triangleright Q$) is called *pure* e-implication (Wiśniewski, 2013, p. 76).

Definition 7 $Q_1 \triangleright Q$ *iff:*

1. *for each direct answer A to the question Q, A entails the disjunction of all the direct answers to the question Q_1, and*

2. *for each direct answer B to the question Q_1 there exists a non-empty proper subset Y of the set of direct answers to the question Q such that B entails the disjunction of all the elements of Y.*

4 Compliance vs pure e-implication

4.1 Translation

In order to compare presented approaches we need to provide a method of interpretation of formulae having the Q-property in INQ in terms of questions in IEL. I will use the method which was presented by Wiśniewski and Leszczyńska-Jasion (2013).

---

[6] A question Q is *sound* iff it has a true direct answer (with respect to the underlying semantics).

I will refer to formulae having the $\mathcal{Q}$-property as Q_{INQ} and to questions in IEL as Q_{IEL}. The procedure is as follows:

1. Compute all the possibilities for a given Q_{INQ}.

2. For each possibility choose exactly one wff such that the possibility is just the truth set of the wff.

3. Each such wff is a possible answer for Q_{IEL}.

Let us consider some examples here.

Example 2 Formula in INQ is: $Q_{\mathrm{INQ}} = (p \vee q) \vee \neg(p \vee q)$. Its set of possibilities is the following: $\lfloor Q_{\mathrm{INQ}} \rfloor = \{|p|, |q|, |\neg p| \cap |\neg q|\}$, thus its IEL counterpart might be formulated as follows: $Q_{\mathrm{IEL}} =?\{p, q, \neg p \wedge \neg q\}$.

Example 3 Formula in INQ is: $Q_{\mathrm{INQ}} = (p \vee \neg p) \wedge (q \vee \neg q)$. Then the set $\lfloor Q_{\mathrm{INQ}} \rfloor = \{|p| \cap |q|, |p| \cap |\neg q|, |\neg p| \cap |q|, |\neg p| \cap |\neg q|\}$ is the set of its possibilities. Thus its IEL counterpart might be formulated as follows: $Q_{\mathrm{IEL}} =?\{(p \wedge q), (p \wedge \neg q), (\neg p \wedge q), (\neg p \wedge \neg q)\}$ (or using abbreviation according to mentioned convention: $Q_{\mathrm{IEL}} =? \pm |p, q|$).

In the following I will use questions in the IEL notation.

4.2 Examples of question responses

Let us now consider some examples of question responses in the light of pure e-implication and compliance.

Example 4 Let us consider the following case. Initiative: $? \pm |p, q|$ and response: $?\{p, \neg p\}$:

- $?\{p, \neg p\} \propto ? \pm |p, q|$
- $?\{p, \neg p\} \triangleright ? \pm |p, q|$

Example 5 Let us take the following initiative: $?\{p \wedge q, p \wedge \neg q, \neg p\}$ and response: $?\{p, \neg p\}$:

- $?\{p, \neg p\} \propto ?\{p \wedge q, p \wedge \neg q, \neg p\}$
- $?\{p, \neg p\} \triangleright ?\{p \wedge q, p \wedge \neg q, \neg p\}$

Example 6 E-implication and compliance hold also for the following initiative: $?\{p, q, \neg p \wedge \neg q\}$ and response: $?\{p \vee q, \neg p \wedge \neg q\}$:

- $?\{p \vee q, \neg p \wedge \neg q\} \propto ?\{p, q, \neg p \wedge \neg q\}$
- $?\{p \vee q, \neg p \wedge \neg q\} \triangleright ?\{p, q, \neg p \wedge \neg q\}$

Example 7 And also for the following initiative: $?\{\neg p, p \wedge q, p \wedge \neg q\}$ and response: $?\{p \rightarrow q, p \wedge \neg q\}$:

- $?\{p \rightarrow q, p \wedge \neg q\} \propto ?\{\neg p, p \wedge q, p \wedge \neg q\}$
- $?\{p \rightarrow q, p \wedge \neg q\} \triangleright ?\{\neg p, p \wedge q, p \wedge \neg q\}$

There are, however, cases where e-implication holds, while the response might not be treated as a compliant one.

Example 8 For the following initiative: $?\{p, \neg p, q, \neg q\}$ and response: $?\{p, \neg p\}$ we have:

- $?\{p, \neg p\} \not\propto ?\{p, \neg p, q, \neg q\}$
- $?\{p, \neg p\} \triangleright ?\{p, \neg p, q, \neg q\}$

Example 9 Similarly for $?\{p, q, \neg p \wedge \neg q\}$ as the initiative, and $?\{p, \neg p\}$ as the response:

- $?\{p, \neg p\} \not\propto ?\{p, q, \neg p \wedge \neg q\}$
- $?\{p, \neg p\} \triangleright ?\{p, q, \neg p \wedge \neg q\}$

Example 10 Both—compliance and e-implication—do not hold in the case when we want the question response to be much more detailed than the initiative, as in the following case.

- $?\{p, \neg p, q, \neg q\} \not\propto ?\{p, \neg p\}$
- $?\{p, \neg p, q, \neg q\} \not\triangleright ?\{p, \neg p\}$

4.3 Compliance is stronger than pure e-implication

Theorem 1 *If $Q_1 \propto Q$ then $Q_1 \triangleright Q$.*

Proof. Suppose that $Q_1 \propto Q$. We should show that both conditions of e-implication for $Q_1 \triangleright Q$ are met (cf. Definition 7).

The first condition for e-implication is met for obvious reasons: as only classical tautologies have $\mathcal{Q}$-property in INQ, Q_1 must be a safe question and thus a sound one (see Wiśniewski, 2013, p. 77, Corollary 7.22).

Now, let us consider the second condition for e-implication, which states that for each direct answer B to the question Q_1 there exists a proper non-empty subset Y of the set of direct answers to the question Q such that B entails the disjunction of all the elements of Y. Compliance demands that for each answer B to the question Q_1 there exists a non-empty subset Y of the set of direct answers to question Q such that the truth set for B is equivalent to the truth set for the disjunction of all formulae in Y. In other words Y is such that B entails the disjunction of all the elements of Y and the disjunction of all the elements in Y entails B. When the stronger condition for compliance will hold also the condition for e-implication will be satisfied. □

We may observe this asymmetry between compliance and e-implication in Example 9. The reason why $?\{p, \neg p\} \not\propto ?\{p, q, \neg p \wedge \neg q\}$ is that there exists the answer to Q_1—namely $\neg p$—for which one cannot point the subset Y, which will meet the first condition for compliance. At the same time e-implication holds because the second condition for e-implication is fulfilled.

When we take a closer look on the Example 8 we will also notice that the second condition of compliance definition (called the restriction clause) makes it stronger than pure e-implication. The intuition behind this clause is that—while proposing Q_1—we cannot rule out a possible answer without providing any information. We may say that the level of information while passing form Q to Q_1 remains the same—the information needed to answer the Q is always enough to answer the Q_1 (cf. Cornelisse, 2009, p. 12). For e-implication it is enough that for each direct answer A to question Q, A entails disjunction of all answers to Q_1. For compliance for each A (which is answer to question Q) there should exist an answer B to question Q_1 such that A entails B. If we consider Example 8 this condition for compliance is not met for the following answers to Q: q and $\neg q$.

5 Summary

In this paper I have introduced approaches to question dependency based in IEL and INQ framework. It is possible to compare compliance and e-implication despite many differences in the background frameworks for these concepts. The obtained result is that compliance is a stronger relationship than pure e-implication. This is the consequence of an intuition behind compliance, saying that a compliant question response should not rule out any possibilities from the initiative. As such compliance might serve as a good source of inspiration for strengthening e-implication.

References

Cornelisse, I. (2009). *A Syntactic Characterization of Compliance in Inquisitive Semantics*. BA Thesis, University of Amsterdam.

Ginzburg, J. (2012). *The Interactive Stance: Meaning for Conversation*. Oxford: Oxford University Press.

Groenendijk, J. (2009). Inquisitive Semantics: Two Possibilities for Disjunction. In P. Bosch, D. Gabelaia, & J. Lang (Eds.), *Proceedings of the 7th International Tbilisi Symposium on Language, Logic, and Computation* (pp. 80–94). Berlin, Heidelberg: Springer.

Groenendijk, J., & Roelofsen, F. (2011). Compliance. In A. Lecomte & S. Tronçon (Eds.), *Ludics, Dialogue and Interaction* (pp. 161–173). Berlin, Heidelberg: Springer.

Łupkowski, P., & Ginzburg, J. (2013). A Corpus-based Taxonomy of Question Responses. In *Proceedings of the 10th International Conference on Computational Semantics (IWCS 2013)* (pp. 354–361). Potsdam: Association for Computational Linguistics.

Urbański, M. (2001). Synthetic Tableaux and Erotetic Search Scenarios: Extension and Extraction. *Logique & Analyse*, *44*, 69–91.

Wiśniewski, A. (1995). *The Posing of Questions: Logical Foundations of Erotetic Inferences*. Dordrecht: Kluwer.

Wiśniewski, A. (2003). Erotetic Search Scenarios. *Synthese*, *134*, 389–427.

Wiśniewski, A. (2013). *Questions, Inferences and Scenarios*. London: College Publications.

Wiśniewski, A., & Leszczyńska-Jasion, D. (2013). *Inferential Erotetic Logic Meets Inquisitive Semantics. Research Report* (Tech. Rep.). Poznań: IntQuestPro.

Paweł Łupkowski
Adam Mickiewicz University
Poland
E-mail: `Pawel.Lupkowski@amu.edu.pl`

Topological Semantics for Conditionals

JOHANNES MARTI
AND RICCARDO PINOSIO

Abstract: In this paper we explore the topological semantics for conditional logic that arises from the Alexandroff equivalence between preorders and topological spaces. This clarifies the relation between the standard order semantics and premise semantics for conditionals. As an application we provide a construction of relative similarity orders between possible worlds from topologies of relevant propositions. The conditional logic over topologies is intertranslatable with the modal logic S4u.

Keywords: conditionals, counterfactuals, similarity orders, premise semantics, evidence models, topological semantics, S4u

1 Introduction

Preorders are the standard semantics for conditional logic. The Alexandroff correspondence associates to every preorder a unique Alexandroff topological space, which is a topological space closed under arbitrary intersections. This suggests using Alexandroff topological spaces as a semantics for conditional logic. We show that this topological semantics captures all that is relevant in premise semantics for the evaluation of conditionals.

We apply this topological approach to the logic of counterfactual conditional. Concretely, we provide a construction of order frames, which are relative similarity orders between worlds, from an Alexandroff topology of relevant propositions. This yields a well-motivated way to obtain a relative similarity order from information which is more basic than similarity among worlds. We completely characterize the class of order frames which results from this construction. This yields strong constraints on order frames which imply the commonly assumed centring condition. We completely axiomatize the validities of this restricted semantics via an intertranslatability result between conditional logic and S4u, which is the usual S4 modal logic on topological spaces, augmented with a universal modality.

2 Alexandroff correspondence for conditionals

2.1 Conditional logic

Various minimization procedures over preorders have naturally arisen in contexts such as the formal semantics of modal notions, belief revision theory or default reasoning (Baltag & Smets, 2006; Imielinski, 1987; Kratzer, 1977; Lewis, 1973). In this paper, we restrict our attention to a setting where a binary modal operator is used to express such a minimization condition. This means we are working with a modal language built according to the following grammar:

$$\varphi ::= p \mid \varphi \wedge \varphi \mid \neg \varphi \mid \varphi \rightsquigarrow \varphi$$

where $p \in \mathsf{Atom}$ is any atomic sentence from a given infinite set Atom. We will call formulas of the form $\varphi \rightsquigarrow \psi$ conditionals. Depending on the application the conditional $\varphi \rightsquigarrow \psi$ can be read in different ways, for instance as "If φ had been the case then ψ would have been the case", "An agent believes ψ conditional on believing φ", or as "If φ then you must ψ".

The semantics for this modal language based on preorders, variously worked out in (Lewis, 1973; Pollock, 1976; Veltman, 1985), has been influential. This approach associates a preorder to every possible world to evaluate the conditional at that world. This yields the following definition:

Definition 1 (Order frame) *An order frame over a set of possible worlds W is a family $(\leq_x)_{x \in W}$, which associates to every world $x \in W$ a preorder (i.e. reflexive, transitive relation) $\leq_x \subseteq W \times W$ over the set W.*

Depending on the setting the preorder associated with a world in an order frame can have different interpretations. In the logic of counterfactual conditionals, the preorders represent metaphysical similarity of the worlds. Thus, $y \leq_x z$ means that the world y is more similar to the world x than the world z. On a doxastic interpretation of the conditional, the preorder associated to a world represents the plausibility of worlds according to the mental state of a given agent at that world. Thus, $y \leq_x z$ means that at x the agent considers it more plausible that y is the actual world than that z is the actual world. We will mostly focus on these two interpretations of the preorder as metaphysical similarity or as plausibility for an agent. It is important to notice that similarity is an ontological notion, while plausibility is an agent relative epistemic notion. For instance for similarity orders, but not for plausibility orders, it is common to require some form of centring,

Topological Semantics for Conditionals

meaning that the world x is in some sense minimal in $\leq_x$. On the other hand for plausibility orders, but not for similarity orders, it is common to require doxastic introspection, meaning that the order associated to any world that the agent considers possible is the same as the preorder in the actual world.

To evaluate formulas on an order frame $(\leq_x)_{x \in W}$ we also need to provide an interpretation for the atomic sentence in Atom. As usual this is done by considering an evaluation function $V\colon \text{Atom} \to \mathcal{P}(W)$ which maps every atomic sentence to the set of worlds where it is true. The semantic clause for the atomic sentences make use of the valuation V, the Boolean connectives are defined as usual, while the semantics of the conditional is as follows:

$$x \models \varphi \rightsquigarrow \psi \quad \text{iff} \quad \text{for all } y \in \llbracket \varphi \rrbracket \text{ there is a } z \in \llbracket \varphi \rrbracket \text{ such that } z \leq_x y$$
$$\text{and } u \models \varphi \to \psi \text{ for all } u \leq_x z.$$

Given a frame $(\leq_x)_{x \in W}$ and a valuation V, we use $\llbracket \cdot \rrbracket$ as a shorthand for the set of all worlds which satisfy the formula φ, i.e. $\llbracket \varphi \rrbracket = \{x \in W \mid x \models \varphi\}$.

It is possible to simplify the semantic clause for the conditional by assuming an additional condition on the frame $(\leq_x)_{x \in W}$ which is called the limit assumption (Warmbrod, 1982). It requires that for any $x \in W$ and set of worlds $X \subseteq W$, X has $\leq_x$-minimal elements. In this case one can check that $x \in \llbracket \varphi \rightsquigarrow \psi \rrbracket$ iff for all the $\leq_x$-minimal elements y in $\llbracket \varphi \rrbracket$ we have that $y \in \llbracket \psi \rrbracket$. It is this fact which explains the notion of minimisation on a preorder we mention above.

The set of validities for the above semantics can be completely axiomatized, see (Veltman, 1985) for details and a completeness proof.

2.2 Alexandroff correspondence

We make use of a correspondence between preorders and topological spaces, first noticed in (Alexandroff, 1937), to derive a topological semantics for the conditional from its order semantics. A topology over a set of points W is family $\tau \subseteq \mathcal{P}(W)$ of sets of points such that $\emptyset, W \in \tau$ and τ is closed under arbitrary unions and finite intersections of sets. The family τ is called the topology of the topological space (W, τ) and its elements are called open sets. We shall just write τ for the topological space (W, τ). The topological spaces that stand in correspondence to preorders are Alexandroff topological spaces, which are additionally closed under arbitrary, not necessarily finite, intersections. We denote an Alexandroff topology by the symbol $\mathcal{A}$ instead of τ to emphasize that the topology is Alexandroff.

We shall now explicitly describe the Alexandroff correspondence. Let $\leq$ be a preorder over over a set W, then the corresponding Alexandroff topology $\mathsf{Do}(\leq) \subseteq \mathcal{P}(W)$ is defined as the set of all downsets of the preorder:

$$\mathsf{Do}(\leq) = \{U \subseteq W \mid \text{if } x \in U \text{ and } y \leq x \text{ then } y \in U\}.$$

In the other direction, for an Alexandroff topology $\mathcal{A} \subseteq \mathcal{P}(W)$ one can define the specialization preorder $\leq^{\mathcal{A}} = \mathsf{Sp}(\mathcal{A})$ on W by:

$$x \leq^{\mathcal{A}} y \quad \text{iff} \quad \text{for all } U \in \mathcal{A} \text{ if } y \in U \text{ then } x \in U.$$

The classic result by Alexandroff is that these constructions establish a bijective correspondence between preorders and Alexandroff spaces.

Theorem 1 *For all preorders $\leq$ on a set W it holds that $\mathsf{Sp}(\mathsf{Do}(\leq)) = \leq$ and for all Alexandroff topologies $\mathcal{A}$ it holds that $\mathsf{Do}(\mathsf{Sp}(\mathcal{A})) = \mathcal{A}$.*

2.3 Topological semantics

The Alexandroff correspondence associates to every preorder an Alexandroff topology; conversely, every Alexandroff topology arises from a unique preorder. We use this fact to obtain a topological semantics for the conditional. We need just replace every preorder $\leq_x$ associated to a world x in an order frame $(\leq_x)_{x \in W}$ with its corresponding Alexandroff topology $\mathsf{Do}(\leq_x)$.

Definition 2 (Neighbourhood space) *A neighbourhood space over a set of possible worlds W is a structure $(\mathcal{A}_x)_{x \in W}$ which associates to each world $x \in W$ an Alexandroff topology $\mathcal{A}_x \subseteq \mathcal{P}(W)$ over the set W.*

By pointwise application of $\mathsf{Do}(\cdot)$ we obtain a neighbourhood space $\mathsf{LDo}((\leq_x)_{x \in W})$ for every order frame $(\leq_x)_{x \in W}$. Analogously, for every neighbourhood space $(\mathcal{A}_x)_{x \in W}$ we obtain an order frame $\mathsf{LSp}((\mathcal{A}_x)_{x \in W})$. This is a bijective correspondence meaning that

- $\mathsf{LSp}(\mathsf{LDo}((\leq_x)_{x \in W})) = (\leq_x)_{x \in W}$ and
- $\mathsf{LDo}(\mathsf{LSp}((\mathcal{A}_x)_{x \in W})) = (\mathcal{A}_x)_{x \in W}$.

To define the topological semantics of conditional logic on a neighbourhood space $(\leq_x)_{x \in W}$ we again need a valuation $V \colon \mathsf{Atom} \to \mathcal{P}(W)$ to fix

Topological Semantics for Conditionals

the truth values of atomic sentences. The semantics for atomic and Boolean formulas is as usual, while for the conditional we have:

$$x \models \varphi \rightsquigarrow \psi \quad \text{iff} \quad \text{for all } \varphi\text{-consistent } U \in \mathcal{A}_x \text{ there is a } \varphi\text{-consistent } V \in \mathcal{A}_x \text{ with } V \subseteq U \text{ and } V \cap [\![\varphi]\!] \subseteq [\![\psi]\!].$$

Again $[\![\varphi]\!] \subseteq W$ is the set of worlds which make the formula φ true and we call a set $U \subseteq W$ φ-consistent if $U \cap [\![\varphi]\!] \neq \emptyset$. The reader can verify that this semantic clause for the conditional on a neighbourhood space corresponds to its semantic clause on the corresponding order frame.

Our topological semantics is akin to the premise semantics for counterfactuals developed in (Kratzer, 1981; Lewis, 1981) and to the evidence models of (van Benthem & Pacuit, 2011). In premise semantics every world in a frame is associated to an arbitrary subset of the powerset of the set of worlds which is not required to be a topology. The elements of this set of sets of worlds associated to a world are called the premises and are thought of as being the relevant background information against which a conditional is evaluated. Lewis (1981) studies the relation between this premise semantics and the order semantics and gives a construction which associates to any order frame a logically equivalent premise frame and viceversa. Lewis does not mention the fact that the construction he uses is the same as the Alexandroff correspondence. Since Lewis does not require the premise sets to be Alexandroff topologies he does not obtain a 1-to-1 correspondence between order frames and premise frames. However, closing a premise set under arbitrary intersections and unions does preserve the truth of all counterfactuals. Therefore one does not collapse any substantial semantic distinctions by requiring premise sets to be Alexandroff topologies, but one gains the 1-to-1 correspondence with order frames.

In the context of epistemic logic, premise frames reappear in the guise of evidence models (van Benthem & Pacuit, 2011). Here the elements of the premise set associated to a world represent a piece of evidence that some agent has about the actual world. Again, Pacuit and van Benthem establish the same relation to order frames but do not mention the underlying Alexandroff correspondence.

3 Coherent neighbourhood spaces and order frames

In this and the following section we discuss an application of the Alexandroff correspondence to the semantics of counterfactual conditional logic.

We observed in the previous section that one can think of the open sets in the local topology $\mathcal{A}_x$ associated to a world x as the facts obtaining at x which are relevant to determine similarity of other worlds to x. The definition of the similarity order $\leq_x$ from the topology $\mathcal{A}_x$ was such that another world y is the more similar to x the more of the relevant propositions in $\mathcal{A}_x$ are true at y. We present a plausible procedure to construct a neighbourhood frame, and hence an order frame, from an independently given Alexandroff topology of relevant propositions. A proposition is deemed to be relevant if its truth or falsity at worlds matters for the relative similarity of worlds. Additionally, we characterize the neighbourhood spaces and order frames that arise in this way, which we name "coherent".

Assume that we are given a set $\mathcal{A}$ of propositions which are relevant to determine the similarity among worlds. For instance one might in general consider the physical laws holding at a world as relevant whereas accidental facts are not relevant. We also assume that $\mathcal{A}$ is an Alexandroff topology, meaning that relevant propositions are closed under arbitrary conjunctions and disjunctions. We mention possible weakenings of this assumption in Section 5. The propositions relevant to determine relative similarity to a particular world x are now all the relevant propositions in $\mathcal{A}$ which are actually true at x. This motivates the following definition.

Definition 3 *For a topological space $\mathcal{A}$ define the corresponding neighbourhood space $\mathsf{Loc}(\mathcal{A}) = (\mathcal{A}_x)_{x \in W}$ by:*

$$\mathcal{A}_x = \{U \in \mathcal{A} \mid x \in U\} \cup \{\emptyset\}.$$

Once we have a neighbourhood space one can apply the Alexandroff correspondence to obtain the corresponding order frame. In this sense Definition 3 provides a construction of relative similarity orders $\mathsf{LSp}(\mathsf{Loc}(\mathcal{A}))$ from any set of relevant propositions $\mathcal{A} \subseteq \mathcal{P}(W)$.

In the rest of this section we characterize the neighbourhood spaces and order frames which arise from an Alexandroff space via this construction.

3.1 Coherent neighbourhood spaces

The neighbourhood spaces arising from the construction above satisfy the following properties:

(C) If $U \in \mathcal{A}_x$, $U \neq \emptyset$ then $x \in U$.

(Ov) If $U \in \mathcal{A}_x$ and $y \in U$, then $U \in \mathcal{A}_y$.

(CU) If $\mathcal{U} \subseteq \bigcup_{x \in W} \mathcal{A}_x$ then $\bigcup \mathcal{U} \in \mathcal{A}_x$ for some x.

We call a neighbourhood space satisfying the above three conditions a coherent neighbourhood space. The first condition (C) is a version of centring for premise frames. The second condition (Ov) captures our assumption that relevance of propositions is not world relative. If a proposition is relevant somewhere then it is relevant wherever it is true.

Theorem 2 *Coherent neighbourhood spaces are in bijective correspondence with Alexandroff topological spaces.*

Proof. The reader can check that the construction from Definition 3 always yields a coherent neighbourhood space.

For the other direction assume we are given any coherent neighbourhood space $(\mathcal{A}_x)_{x \in W}$. We define the following Alexandroff topology on W:

$$\mathsf{Col}((\mathcal{A}_x)_{x \in W}) = \bigcup_{x \in W} \mathcal{A}_x.$$

Using the conditions on coherent neighbourhood spaces one can check that $\bigcup_{x \in W} \mathcal{A}_x$ is indeed an Alexandroff topology.

It is easy to check that $\mathsf{Col}(\mathsf{Loc}(\mathcal{A})) = \mathcal{A}$ for any Alexandroff space $\mathcal{A}$. For the other direction, we show that $(\mathcal{A}'_x)_{x \in W} = \mathsf{Loc}(\mathsf{Col}((\mathcal{A}_x)_{x \in W}))$ is equal to $(\mathcal{A}_x)_{x \in W}$ for any neighbourhood space $(\mathcal{A}_x)_{x \in W}$. We need that $\mathcal{A}_x = \mathcal{A}'_x$ for all $x \in W$. So take any $U \in \mathcal{A}_x$. Then $U \in \mathsf{Col}((\mathcal{A}_x)_{x \in W})$. Hence it is also in $\mathcal{A}'_x = \{U \in \mathsf{Col}((\mathcal{A}_x)_{x \in W}) \mid x \in U\} \cup \{\emptyset\}$ because by (C) it follows from $U \in \mathcal{A}_x$ that $x \in U$. If on the other hand we have $U \in \mathcal{A}'_x = \{U \in \bigcup_{x \in W} \mathcal{A}_x \mid x \in U\} \cup \{\emptyset\}$ then U is either empty hence clearly $U \in \mathcal{A}_x$ or there is some $y \in W$ such that $U \in \mathcal{A}_y$ and $x \in U$. By (Ov) it follows that $U \in \mathcal{A}_x$. $\square$

3.2 Coherent order frames

Corresponding to the notion of a coherent neighbourhood space we define coherent order frames as the class of order frames $(\mathcal{A}_x)_{x \in W}$ which satisfy the following two axioms:

(Ex) For all $x, y, z \in W$: If $x \leq_y z$ then $x \leq_z y$.

(St) For all $x, y, z \in W$: If $x \leq_y z$ then $x \leq_y y$ or $x \leq_z z$.

A version of centring is implied by (Ex): Because by reflexivity $x \leq_y x$ it follows that $x \leq_x y$ for all $x, y \in W$. This version of centring is slightly weaker than the strong centring which requires $x \leq_x y$ and not $y \leq_x x$ for all $x, y \in W$. It is, however, stronger than weak centring which requires only hat x is a minimal element in the order $\leq_x$. For discussion of different centring conditions see (Veltman, 1985).

The two conditions (Ex) and (St) are quite strong. One can check that neither is satisfied by relative closeness of points in a metric space, as considered e.g. in (Lehmann, Magidor, & Schlechta, 2001). However, the two conditions correspond exactly to the coherence conditions on neighbourhood spaces, which are justified by the intuitive construction of a neighbourhood space from the set of relevant propositions.

Theorem 3 *Coherent order frames are in bijective correspondence with coherent neighbourhood spaces.*

We first need some lemmas which require an additional notion. Define the minimal neighbourhood of y in the topology $\mathcal{A}_x$ of a neighbourhood space $(\mathcal{A}_x)_{x \in W}$ by:

$$N_x(y) = \bigcap \{U \in \mathcal{A}_x \mid y \in U\}.$$

Since $\mathcal{A}_x$ is an Alexandroff topology $N_x(y)$ is again open and it is in fact the minimal open set in $\mathcal{A}_x$ containing y.

Lemma 1 *Let $(\leq_x)_{x \in W}$ be an order frame. Then $(\leq_x)_{x \in W}$ satisfies (Ex) if and only if $\mathsf{LDo}((\leq_x)_{x \in W})$ satisfies for all $y, z \in W$*

(Ex') $N_y(z) = N_z(y)$

and it satisfies (St) if and only if $\mathsf{LDo}((\leq_x)_{x \in W})$ satisfies for all $y, z \in W$

(St') $N_y(z) \subseteq N_y(y) \cup N_z(z)$.

Proof. Omitted. □

Lemma 2 *For every neighbourhood space $(\mathcal{A}_x)_{x \in W}$ we have:*

1. *(C) and (Ov) imply (Ex').*

2. *(Ov) and (CU) imply (St').*

3. *(Ex') implies (C).*

Topological Semantics for Conditionals

4. (Ex') and (St') imply (Ov).

5. (St') and (Ov) imply (CU).

Proof. We leave the points 1 and 3 to the reader.

For point 2: By (CU), $N_y(y) \cup N_z(z) \in \mathcal{A}_w$ for some w. But then because $y \in N_y(y)$ we have by (Ov) that $N_y(y) \cup N_z(z)$ is in $\mathcal{A}_y$. Because $z \in N_z(z) \subseteq N_y(y) \cup N_z(z)$ and $N_y(z)$ is the minimal open set in $\mathcal{A}_y$ containing z it follows that $N_y(z) \subseteq N_y(y) \cup N_z(z)$.

For point 4: consider $U \in \mathcal{A}_x$ and $y \in U$. To show that $U \in \mathcal{A}_y$, we show that for any $z \in U$, there exists an open set $O \in \mathcal{A}_y$ such that $z \in O$ and $O \subseteq U$. Let $z \in U$, and consider $N_y(z)$: it is open in $\mathcal{A}_y$, and z belongs to it. We show that $N_y(z) \subseteq U$. By (St'), $N_y(z) \subseteq N_y(y) \cup N_z(z)$. Hence it is w.l.o.g. enough to show that $N_y(y) \subseteq U$:

$$
\begin{aligned}
N_y(y) &\subseteq N_y(x) & & y \in N_y(x) \text{ by (C)} \\
&= N_x(y) & & \text{(Ex')} \\
&\subseteq U & & y \in U \in \mathcal{A}_x
\end{aligned}
$$

For point 5: consider a family of sets $\mathcal{U} \subseteq \bigcup_{x \in W} \mathcal{A}_x$. If $\bigcup \mathcal{U} = \emptyset$, then certainly $\bigcup \mathcal{U} \in \mathcal{A}_x$ for any x. Otherwise, pick $B \in \mathcal{U}$ non-empty, and point $b \in B$. We prove that $\bigcup \mathcal{U} \in \mathcal{A}_b$. It suffices to show that for any $x \in \bigcup \mathcal{U}$ the set $N_b(x)$ is open in $\mathcal{A}_b$, and $x \in N_b(x) \subseteq \bigcup \mathcal{U}$. By definition $N_b(x)$ is open in $\mathcal{A}_b$ and $x \in N_b(x)$. To show that $N_b(x) \subseteq \bigcup \mathcal{U}$ it is enough to prove that $N_b(b) \subseteq \bigcup \mathcal{U}$ and that $N_x(x) \subseteq \bigcup \mathcal{U}$, because by (St'), $N_b(x) \subseteq N_b(b) \cup N_x(x)$. The former holds because $N_b(b) \subseteq B$, since $b \in B \in \mathcal{A}_b$, and $B \subseteq \bigcup \mathcal{U}$, since $B \in \mathcal{U}$. For the latter take a $U \in \mathcal{U}$ such that $x \in U$. By (Ov) it follows that $U \in \mathcal{A}_x$ and hence $N_x(x) \subseteq U \subseteq \bigcup \mathcal{U}$. □

We can now prove Theorem 3.

Proof. (of Theorem 3) Let $(\mathcal{A}_x)_{x \in W}$ be a coherent neighbourhood space. By 1. and 2. of Lemma 2 it must satisfy conditions (Ex') and (St') of Lemma 1. Hence, the corresponding order frame $\mathsf{LSp}((\mathcal{A}_x)_{x \in W})$ is coherent.

Let $(\leq_x)_{x \in W}$ be a coherent order frame. By Lemma 1 the corresponding neighbourhood space $\mathsf{LDo}((\leq_x)_{x \in W})$ satisfies (Ex') and (St'). By points 3. 4. and 5. of Lemma 2 it follows that $\mathsf{LDo}((\leq_x)_{x \in W})$ is a coherent neighbourhood space.

The correspondence between order frames and neighbourhood spaces remains bijective when restricted to coherent order frames and coherent neighbourhood spaces. □

4 S4u and completeness

The purpose of the present section is to show that the conditional logic over coherent neighbourhood spaces and S4u over topological spaces are inter-translatable. This gives us an easy route to the completeness of conditional logic on coherent neighbourhood spaces. We obtain the complete axiomatization by translating the already know axiomatization of S4u into the language of the conditional and by adding axioms that guarantee the provability of the translation rules.

The results of this section also hold for coherent neighbourhood spaces in which the local topologies associated to a world are not required to be Alexandroff. Such a space can be obtained from an arbitrary topological space of relevant propositions by means of the localisation procedure described in Definition 3. In the following we however stick to the notation $\mathcal{A}$ and $\mathcal{A}_x$ for the topology of a space and for the local topology at a world x.

We give a brief overview of the aspects of S4u that we need here. For further details consult (Aiello, van Benthem, & Bezhanishvili, 2003). The language of S4u is a bimodal language with two unary modalities □ and ∀. On a topological space the semantics of □ is the interior modality whereas ∀ is a universal modality. We formulate the semantics on the level of the coherent neighbourhood space generated from a topological space. By Theorem 2 this is the same as working with just topological spaces. The most convenient formulation of the semantics for our purposes is:

$$x \models \forall \varphi \quad \text{iff} \quad U \subseteq [\![\varphi]\!] \text{ for all } U \in \mathcal{A}_x.$$
$$x \models \Box \varphi \quad \text{iff} \quad U \subseteq [\![\varphi]\!] \text{ for some non empty } U \in \mathcal{A}_x.$$

Note that this induces the following clauses for the duals:

$$x \models \exists \varphi \quad \text{iff} \quad \text{some } U \in \mathcal{A}_x \text{ is } \varphi\text{-consistent}.$$
$$x \models \Diamond \varphi \quad \text{iff} \quad \text{all non-empty } U \in \mathcal{A}_x \text{ are } \varphi\text{-consistent}.$$

The validities of this semantics are axiomatized by the modal logic S4u in which □ is an S4 modality, ∀ an S5 modality and the interaction axiom $\forall \varphi \to \Box \varphi$ holds.

Topological Semantics for Conditionals

The intuitive reading of the universal $\forall$ modality is metaphysical necessity. This also becomes clear in the first translation rule of the following theorem, which amounts to the same embedding of metaphysical necessity into conditional logic as suggested in appendix 1 of (Williamson, 2007):

Theorem 4 *These equivalences hold on coherent neighbourhood spaces:*

$$\forall \varphi \equiv \neg \varphi \rightsquigarrow \bot$$
$$\Box \varphi \equiv \top \rightsquigarrow \varphi$$
$$\varphi \rightsquigarrow \psi \equiv (\Diamond \varphi \rightarrow \Box(\varphi \rightarrow \psi)) \wedge (\neg \Diamond \varphi \rightarrow \forall(\varphi \rightarrow \Diamond(\varphi \wedge \Box(\varphi \rightarrow \psi))))$$

Proof. The proof of the first two equivalences is omitted.

For the third equivalence we start with the left to right direction. Let $(\mathcal{A}_x)_{x \in W}$ be a coherent neighbourhood space and $x \in W$ such that $x \models \varphi \rightsquigarrow \psi$. We distinguish two cases. Either all non empty opens in $\mathcal{A}_x$ are φ-consistent or there is a non empty $U \in \mathcal{A}_x$ which is not φ-consistent.

In the first case where they are all φ-consistent it follows that $x \models \Diamond \varphi$. To show $x \models \Box(\varphi \rightarrow \psi)$ pick any non empty $U \in \mathcal{A}_x$. We can assume that U is φ-consistent, because otherwise U witnesses the truth of $\Box(\varphi \rightarrow \psi)$. It follows by $x \models \varphi \rightsquigarrow \psi$ that there is a $V \in \mathcal{A}_x$ such that $V \cap \llbracket \varphi \rrbracket \subseteq \llbracket \psi \rrbracket$. So V witnesses the truth of $\Box(\varphi \rightarrow \psi)$.

In the second case, where there is a non empty $U \in \mathcal{A}_x$ that is not φ-consistent, it follows that $x \models \neg \Diamond \varphi$. We need to show that $x \models \forall(\varphi \rightarrow \Diamond(\varphi \wedge \Box(\varphi \rightarrow \psi)))$. To prove this it is sufficient to take an arbitrary $v \in \llbracket \varphi \rrbracket$ and to show that $v \models \Diamond(\varphi \wedge \Box(\varphi \rightarrow \psi))$. Pick any non empty $V \in \mathcal{A}_v$. We need the existence of a $z \in V$ such that $z \models \varphi \wedge \Box(\varphi \rightarrow \psi)$.

Consider $U \cup V$. Since $U \in \mathcal{A}_x$ and $V \in \mathcal{A}_v$, then it follows by (CU) that $U \cup V \in \mathcal{A}_w$ for some $w \in W$. Because $U \in \mathcal{A}_x$ it follows by (C) that $x \in U \subseteq U \cup V$. So it follows by (Ov) that $U \cup V \in \mathcal{A}_x$. Also because $V \in \mathcal{A}_v$ it follows again by (C) that $v \in V \subseteq U \cup V$. So $U \cup V$ is φ-consistent because $v \in \llbracket \varphi \rrbracket$. Now we use the assumption that $x \models \varphi \rightsquigarrow \psi$ to obtain a φ-consistent $Z \in \mathcal{A}_x$ with $Z \subseteq U \cup V$ and $Z \cap \llbracket \varphi \rrbracket \subseteq \llbracket \psi \rrbracket$. Pick a $z \in Z$ which satisfies φ. Then $z \in V$ because $Z \subseteq U \cup V$ and U is assumed to be not φ-consistent. Now we can see that $z \models \Box(\varphi \rightarrow \psi)$ since $Z \subseteq \llbracket \varphi \rightarrow \psi \rrbracket$, Z is not empty, and $Z \in \mathcal{A}_z$ by (Ov).

For the right to left direction suppose that the formula on the right is true at some world x. We need to show that then $x \models \varphi \rightsquigarrow \psi$. Again, we distinguish the cases where all non empty elements of $\mathcal{A}_x$ are φ-consistent and where there is an element in $\mathcal{A}_x$ that is not φ-consistent.

If all non empty $U \in \mathcal{A}_x$ are φ-consistent then $x \models \Diamond\varphi$ and by assumption we also have that $x \models \Box(\varphi \to \psi)$. Hence there is a non empty $Z \in \mathcal{A}_x$ with $Z \subseteq [\![\varphi \to \psi]\!]$. To prove that $x \models \varphi \leadsto \psi$ pick any φ-consistent $U \in \mathcal{A}_x$. Now consider the intersection $U \cap Z \subseteq U$. It is in $\mathcal{A}_x$ since $\mathcal{A}_x$ is a topology. By (C) we have that $x \in U \cap Z$ hence it is a non empty element of $\mathcal{A}_x$. Since we are in the case where all non empty elements of $\mathcal{A}_x$ are φ-consistent it follows that $U \cap Z$ is also φ-consistent. It holds that $U \cap Z \cap [\![\varphi]\!] \subseteq [\![\psi]\!]$ because $U \cap Z \subseteq Z \subseteq [\![\varphi \to \psi]\!]$.

In the other case there is a non empty $Z \in \mathcal{A}_x$ which is not φ-consistent. Then $x \models \neg\Diamond\varphi$ and so $x \models \forall(\varphi \to \Diamond(\varphi \wedge \Box(\varphi \to \psi)))$. In order to show that $x \models \varphi \leadsto \psi$ take any φ-consistent $U \in \mathcal{A}_x$. Because it is φ-consistent there is a $u \in U$ with $u \models \varphi$. So U is a non-empty element of $\mathcal{A}_x$ and therefore by $x \models \forall(\varphi \to \Diamond(\varphi \wedge \Box(\varphi \to \psi)))$ we have that $U \subseteq [\![\varphi \to \Diamond(\varphi \wedge \Box(\varphi \to \psi))]\!]$. So $u \models \Diamond(\varphi \wedge \Box(\varphi \to \psi))$. Because of (Ov) we have $U \in \mathcal{A}_u$ and so U is $\varphi \wedge \Box(\varphi \to \psi)$-consistent. Hence there is a $v \in U$ with $v \models \varphi$ and $v \models \Box(\varphi \to \psi)$. By the latter it follows that there is a non-empty $V \in \mathcal{A}_v$ such that $V \subseteq [\![\varphi \to \psi]\!]$. By (CU) $V \cup Z \in \mathcal{A}_w$ for some world W. Because $x \in Z$ it follows that $V \cup Z \in \mathcal{A}_x$ by (Ov). Since $\mathcal{A}_x$ is closed under intersections we then obtain $(V \cup Z) \cap U \in \mathcal{A}_x$. Clearly $(V \cup Z) \cap U \subseteq U$ and $(V \cup Z) \cap U \subseteq V \cup Z \subseteq [\![\varphi \to \psi]\!] \cup [\![\neg\varphi]\!] \subseteq [\![\varphi \to \psi]\!]$ because Z is not φ-consistent. $(V \cup Z) \cap U$ is φ-consistent because $v \in V$ by (C) and hence $v \in (V \cup Z) \cap U$. $\square$

Corollary 1 *The validities of conditional logic over coherent neighbourhood spaces are completely axiomatizable.*

Proof. One can obtain an axiomatization by translating the S4u rules and axioms into the language of the conditional using the translation clauses in Theorem 4 and adding the translation of the third equivalence from Theorem 4 as an axiom. $\square$

5 Conclusions and further work

In this paper we have done the following. We have used the Alexandroff correspondence between preorders and Alexandroff topological spaces to:

- clarify the formal relation between the order semantics for conditionals and its semantics based on sets of sets of worlds, as in the case of premise frames and evidence models.

- present a topological semantics for conditional logic.

- provide a construction to generate a similarity order among worlds starting from a set of relevant propositions. We also characterized the order frames that arise from this construction.

Additionally, we established the intertranslatability of the logic of counterfactual over topological spaces and S4u, by means of which we obtained a completeness result.

Some possible directions for further work are:

- One might try to weaken the somewhat implausible assumption that the set of relevant propositions is closed under disjunctions. This might lead to a weaker notion of coherence for neighbourhood spaces and order frames.

- One can try to abstract away from possible worlds and construct them and the relative similarity order from an algebra of propositions.

- The axiomatization of conditional logic on coherent order frames obtained in Corollary 1 is not aesthetically pleasing. It remains an open question whether there are more natural axioms.

- Our construction of a relative similarity order from a set of relevant propositions treats all relevant propositions as equally relevant. However, it seems natural to rank propositions according to their importance, for instance see (Lewis, 1979, p. 472). This ranking might be implemented by an ordering over the set of relevant propositions.

We hope to address some of these points in a future paper.

References

Aiello, M., van Benthem, J., & Bezhanishvili, G. (2003). Reasoning About Space: The Modal Way. *Journal of Logic and Computation*, *13*, 889–920.

Alexandroff, P. (1937). Diskrete Räume. *Mat. Sbornik (NS)*, *2*, 501–518.

Baltag, A., & Smets, S. (2006). Conditional Doxastic Models: A Qualitative Approach to Dynamic Belief Revision. *Electronic Notes in Theoretical Computer Science*, *165*, 5–21.

Imielinski, T. (1987). Results on Translating Defaults to Circumscription. *Artificial Intelligence*, *32*, 131–146.

Kratzer, A. (1977). What 'Must' and 'Can' Must and Can Mean. *Linguistics and Philosophy*, *1*, 337–355.

Kratzer, A. (1981). Partition and Revision: The Semantics of Counterfactuals. *Journal of Philosophical Logic*, *10*, 201–216.

Lehmann, D. J., Magidor, M., & Schlechta, K. (2001). Distance Semantics for Belief Revision. *Journal of Symbolic Logic*, *66*, 295-317.

Lewis, D. K. (1973). *Counterfactuals*. Oxford: Blackwell.

Lewis, D. K. (1979). Counterfactual Dependence and Time's Arrow. *Noûs*, *13*, 455–476.

Lewis, D. K. (1981). Ordering Semantics and Premise Semantics for Counterfactuals. *Journal of Philosophical Logic*, *10*, 217–234.

Pollock, J. L. (1976). The 'Possible Worlds' Analysis of Counterfactuals. *Philosophical Studies*, *29*, 469–476.

van Benthem, J., & Pacuit, E. (2011). Dynamic Logics of Evidence-Based Beliefs. *Studia Logica*, *99*, 61–92.

Veltman, F. (1985). *Logics for Conditionals*. PhD Thesis, University of Amsterdam.

Warmbrod, K. (1982). A Defense of the Limit Assumption. *Philosophical Studies*, *42*, 53–66.

Williamson, T. (2007). *The Philosophy of Philosophy*. Malden, MA: Blackwell.

Johannes Marti
ILLC, University of Amsterdam
The Netherlands
E-mail: johannes.marti@gmail.com

Riccardo Pinosio
ILLC, University of Amsterdam
The Netherlands
E-mail: rpinosio@gmail.com

Modal Validity and the Dispensability of the Actuality Operator

VITTORIO MORATO

Abstract: In this paper, I claim that two ways of defining validity for modal languages ("real-world" and "general" validity), corresponding to distinction between a correct and an incorrect way of defining modal validity, correspond instead to two substantive ways of conceiving modal truth. At the same time, I claim that the major logical manifestation of the real-world/general validity distinction in modal propositional languages with the actuality operator should not be taken seriously, but simply as a by-product of the way in which the semantics of such an operator is usually given.

Keywords: real-world validity, general validity, actuality operator, propositional modal logics

1 Introduction

Take two modal logicians and call them "Saul" and "Max". There is some chance that what Saul and Max mean by "validity" of a formula of a modal language (with respect to a certain class of interpretations) is different. What Saul might mean is that a formula ϕ of a modal language L is valid, with respect to a certain class of interpretations, iff ϕ is true in *every actual world* of every interpretation of L; what Max might mean instead is that ϕ is valid in L iff ϕ is true in *every world* of every interpretation of L. According to Max, validity for a modal language is some sort of "super-necessity"; it is what remains necessary under every interpretation. According to Saul, validity is some sort of "super-actuality"; it is what remains (actually) true under every interpretation. For Saul, modal validity is "permuted" actuality, for Max, validity is "permuted" necessity.[1]

It could be claimed that the disagreement between Saul and Max might be resolved by invoking a pluralistic approach to logic. A logical pluralist

---

[1] The names "Saul" and "Max" have not been chosen arbitrarily. Saul is inspired by Saul Kripke, who introduced the first way of defining modal validity, for example in (Kripke, 1963a); Max is inspired by Max Cresswell, who usually defines modal validity in the second way (Hughes & Cresswell, 1996). In another paper of mine (Morato, 2014) I have called "Kripke validity" the former notion and "Textbook validity" the latter.

would react basically by denying the reality of this disagreement and by conceding that the difference between (what has been called) "real-world validity" and "general validity" is not, after all, to be taken seriously. The entire issue could be resolved by an appropriate disambiguation of the term "validity".[2]

My aim in this paper is twofold. On the one hand, I will show that the difference between the two notions is to be taken seriously, because it corresponds to a real, substantive distinction between two general conceptions of modality. On the other hand, I will show that the difference should not be motivated, as it is usually done, by appealing to modal propositional languages enriched with the actuality operator.

This might sound bizarre: what I will do, in effect, is to defend the plausibility and robustness of a (meta)logical distinction and to dismiss, at the same time, one of its major logical manifestations. The moral of this situation will be discussed in the final section of the paper.

2 On the genealogy of modal validity

In this section, I will show how the distinction between real-world validity (truth in every actual world of every interpretation) and general validity (truth in every world of every interpretation), far from being a distinction between a correct and an incorrect definition of validity for modal languages, could be grounded on the difference between two general conceptions of modality.[3]

In order for these two general conceptions to be made explicit, I will avail myself of a certain view on how modal logic emerges from its non-modal basis.[4]

Before doing this, however, some preliminaries are necessary. An interpretation of modal propositional logic L is a quadruple, $\langle W, R, @, V \rangle$,

---

[2]The kind of logical pluralism I have in mind is the one defended for example in (Beall & Restall, 2006). The "real-world"/"general" validity terminology comes from (Davies & Humberstone, 1980). Even though I will use such terminology for the rest of the paper, I am not completely happy with it, because I think it is more suited for discussions about validity in two-dimensional modal logics. (Morato, 2014) is more directly related to this topic.

[3](Nelson & Zalta, 2012; Zalta, 1988) defend the view that general validity is not a "correct" definition of validity, because it is based on a conflation of a semantic notion (truth in an interpretation) with a metaphysical one (necessity) and because it is not in full compliance with the Tarskian approach to logical truth. I rebut both claims in (Morato, 2014).

[4]This view is inspired by and is very similar to the one presented in (Menzel, 1990).

where W is an arbitrary set of objects, @ is an element of W, R is a total, binary relation on W and V is a valuation function that assigns to each propositional atom, with respect to an element of W, precisely one element of the set $\{0, 1\}$. In the intended interpretation of L, the set W is a set of worlds, @ is the actual world, V is a function that tells us which atomic sentences are true in what world and R is a relation of accessibility among worlds. The basic metalogical notion of "truth of a formula ϕ in an interpretation I of L with respect to a world w" is defined recursively in the usual way. Other metalogical notions are "truth in an interpretation I" and "validity" (truth in all interpretations). If one defines a frame F as the pair formed by $\langle W, R \rangle$, one could define further metalogical notions such as "truth in w in a frame F", "validity in a frame F" and "validity in a class of frames".[5] The difference between real-world and general validity emerges in case one decides to use @ in the definition of the notion of truth in an interpretation I: those who decide to define such a notion as truth in the @ of I (treating truth in an interpretation as actual truth) will have, as a result, that the notion of validity (truth in all interpretations) is truth in all actual worlds of all interpretations. Those who decide to define the notion of truth in an interpretation as truth in every world of an interpretation (treating truth in an interpretation as necessary truth), will have, as a result, that the notion of validity is truth in every world of every interpretation. Those belonging to this latter group usually do not include @ in the definition of an interpretation.[6]

Now, all of this could be fruitfully recarved in the following way. Modal propositional logic could be seen as a *generalization* of non-modal propositional logic. While non-modal propositional truth is defined over a single interpretation—where an interpretation of propositional logic is $\langle V \rangle$ and V is an assignment of truth-values to propositional atoms—modal propositional truth is defined over a *cluster* of non-modal propositional interpretations. As far as formal semantics is concerned thus, the role of the elements in W might simply be viewed as that of *indexing* a cluster of non-modal propositional interpretations. Modal operators could then be seen as operators that quantify, in the metalanguage, over indexed non-modal propositional interpretations. A formula such as $\Box \phi$ is thus true if ϕ is true in the relevant set of indexed propositional non-modal interpretations, $\Diamond \phi$ is true

---

[5] See (Beall & Restall, 2006; Chellas, 1980).

[6] The specification of @ in a modal interpretation might be avoided if one defines validity as truth in every possible world in every interpretation, but it is essential if one wants to introduce an actuality operator defined by a clause like (2) (see page 134).

if ϕ is true in at least one relevant indexed propositional non-modal interpretation.

Given this picture, modal truth might be viewed as emerging from non-modal truth depending on how the set of indexed propositional interpretations is generated. There are two ways in which this could happen and these two ways correspond, in my view, to two ways in which the space of possibilities can be conceived.

According to the first method, in order to generate such a cluster of non-modal interpretations, you first take a specific non-modal propositional interpretation, that we could indicate with '$I^@$', and associate to $I^@$ a cluster of *alternative* non-modal propositional interpretations. These alternative non-modal interpretations could be called the "variants" of $I^@$ and are to be indexed by the members of the set W. Under this picture, a modal propositional interpretation is obtained by the association of $I^@$ with its variants. The original $I^@$ will play the role of the actual world and the variants of $I^@$ will play the role of possible worlds. What is true according to $I^@$ is true *simpliciter*, what is true according to a variant of $I^@$ is possibly true.

As far as metalogical notions are concerned, the natural choice to do, in this case, is to characterize the notion of truth in a modal interpretation (truth in I of L) as truth in the $I^@$ of the interpretation or, in other terms, as truth in the actual world of the interpretation. The notion of validity will be thus characterized as truth in every $I^@$ of every modal propositional interpretation. Given that $I^@$ is simply a non-modal propositional interpretation, the definition of modal validity and of non-modal validity will quantify over the same set of interpretations.

The second method is more straightforward. According to this method, a modal propositional interpretation is simply formed by a certain number of indexed non-modal propositional interpretations; there is no "privileged" interpretation and every non-modal propositional interpretation is, semantically, on a par.

Under this picture, the natural choice to do, as far as modal metalogical notions are concerned, is to *downgrade* non-modal meta-logical notions. Given that modal propositional logic is a generalization of non-modal propositional logic, a non-modal metalogical notion of level n will be a modal metalogical notion of level $n-1$. Non-modal truth in an interpretation will now be modal truth in an interpretation with respect to an indexed interpretation: non-modal propositional truth in I will be downgraded to the basic modal notion of "truth in w in I". Non-modal validity, namely truth in every non-modal propositional interpretation will be downgraded to modal truth in a single modal propositional interpretation: modal propositional validity is truth in every indexed non-modal propositional interpretation, namely truth in every possible world.

These two methods correspond to two different views on the nature of modal space. The first method corresponds to a view according to which the space of possibilities is seen *sub specie actualitatis*, while the second method corresponds to a view according to which the space of possibilities is seen *sub specie possibilitatis*. I will now try to clarify these two general conceptions.

According to the first conception, actuality has a privileged status over possibilities. In particular, what possibilities there are depends on what is actual. The logical space of possibilities is determined by the ways the (actual) world might have been. This idea is represented, in the first method, by the fact that possible worlds are treated as indexed *variants* of $I^@$. A consequence of an actuality-constrained conception of possibility is that possible truth does not "interfere" with plain truth: this is equally represented in the first method by the fact that what is true in a modal interpretation is what is true at the $I^@$ of the interpretation and by the fact that, in order to determine what is valid, only what is true in the $I^@$ of every interpretation is relevant. Only what is actually true in every interpretation contributes to validity.

According to the second conception, actuality has not a privileged status over possibilities. In this case, the logical space of possibilities is not constrained by what is actual, in the sense that there could be some "free-floating" (or "alien") possibilities, i.e., possibilities that are not to be conceived as simply false descriptions of the actual world. A consequence of this non-constrained conception of possibility is that possible truth and actual truth are on a par: this idea is represented in the second method by the fact that what is true in a modal interpretation is what remains true in every indexed non-modal interpretation and by the fact that, in order to determine what is valid, every truth in every indexed non-modal interpretation is relevant; it is what is necessarily true in every interpretation that contributes to validity.

As far as simple propositional modal languages are concerned (modal propositional languages with only $\Box$ and $\Diamond$), the two conceptions are extensionally equivalent: for simple modal propositional languages, one who conceives modal space *sub specie actualitatis* and one who conceives modal space *sub specie possibilitatis* will take exactly the same formulas as valid. In terms of the cluster conception developed above, this is easily seen by considering that every non-modal propositional interpretation will happen to be the privileged $I^@$ of a modal propositional interpretation; if this is so, what is true in every $I^@$ of every interpretation *just is* what is true in every indexed interpretation.

Real-world validity and general validity are thus extensionally equivalent for simple modal languages. The recarving strategy and the two conceptions of modality I have presented in this section might be taken as ways in which one can see how these two notions, while extensionally equivalent, are intensionally different.

3 The actuality operator enters the scene

In modal languages enriched with the actuality operator A, there seems to be formulas that are valid, if the notion of real-world validity is used and invalid, if the notion of general validity is used instead. In such languages, the intensional difference between the two notions seems, in effect, to make an extensional difference.

The actuality operator has been introduced in modal languages in order to satisfy the expressive need of breaking the scope of modal operators. This need is especially pressing in the first-order case, where we need to express the formal counterparts of English sentences such as:

It could happen that all those that are actually rich should be poor

where the quantifier binding those actually rich should be interpreted outside the scope of the initial modal operator 'it could happen that'. The translation of this sentence, in a first-order modal language, is the following:

$$\Diamond \forall x (\mathrm{A} Rx \to Px) \qquad (1)$$

The interpretation of such a formula depends, of course, on the semantics clause for A and on the interactions between quantification and modality. The semantics clause for A is usually given by the following clause:

$$\mathrm{A}\phi \text{ is true in } L^{\mathrm{A}} \text{ iff } V(\phi, @) = 1 \qquad (2)$$

where L^A is a standard modal propositional language enriched with the actuality operator. The main effect of (2) is that of making A a tool to obtain *truth-value rigidity*: when applied to a formula ϕ, A sticks to ϕ the truth-value that ϕ has in the actual world and Aϕ designates the truth-value that ϕ has in the actual world in every possible world.

It is rarely noticed that, even armed with (2), the intended reading of formula (1) is not easily captured in first-order modal logic: in particular, in a system with variable domains and actualist quantification[7], for a formula such as (1) to come out true, it is enough that there exists an interpretation I, where there is at least one world w such that no individual in w exists also in the @ of I. The intended reading of formula (1) thus can only be captured in a system with fixed or increasing domains.

If we define A by means of (2), the following formula:

$$\phi \leftrightarrow A\phi \qquad (3)$$

will be real-world valid, but not generally valid.

To prove that a formula is real-world valid but not generally valid, it is enough to prove that the formula is real-world valid, but not necessary. To prove that (3) is real-world valid, assume first that ϕ is true in the @ of an arbitrary interpretation I of a modal propositional language L^A. Given (2) (right-to-left), even $A\phi$ will be true in @ and thus $\phi \to A\phi$ is real-world valid. If $A\phi$ is true in the @ of an arbitrary interpretation I, then again by (2) (left-to-right), ϕ will be true in @ and thus $A\phi \to \phi$ is real-world valid. To prove that (3) is not necessary (and then not generally valid) it is enough to show that $\phi \to A\phi$ is not necessary. Consider the following interpretation:

- $w = \{w_1, w_2, w_3\}$
- @ $= w_1$
- $V(w_1, \phi) = 0, V(w_2, \phi) = 1, V(w_3, \phi) = 1$

In w_2, the antecedent is true, but the consequent is false, and so there is at least a world where the conditional is false. Not being necessary, the formula cannot be generally valid. (3) is thus a case of a formula that is real-world valid without being generally valid.

The fact that (3) is real-world valid without being necessary implies that the rule of necessitation (RN) fails for non-simple modal languages with real-world validity. The difference between real-world validity and general validity could be thus reframed as a difference with respect to RN: general and real-world validity differ, because RN preserves the former, but it does not preserve the latter.

---

[7]The famous system of (Kripke, 1963b), presented also in Chapter 15 of (Hughes & Cresswell, 1996).

It might be interesting to understand what types of formulas are real-world valid without being generally valid. Given the equivalence of real-world and general validity for simple modal languages, we know that such formulas will be formulas with at least one occurrence of A. Call a formula with at least one occurrence of A an A-formula. The problem thus is to determine what kind of A-formulas are real-world valid without being necessary (and thus, *mutatis mutandis*, generally valid).

Given (2), a real-world valid A-formula governed by the actuality operator (a formula of the form $A\phi$) is also generally valid. If $A\phi$ is real-world valid, it is true in the actual world of every interpretation; consider such an interpretation I; in I, $A\phi$ is true in the @ of I. By (2), it follows that ϕ is true in the @ of I. But if ϕ is true in the @ of I, in every world w of I it will be true that $A\phi$; $A\phi$ will be necessary and thus, *mutatis mutandis*, generally valid.

What about formulas governed by $\Diamond$ and $\Box$? Here the question is whether real-world valid A-formulas of the form $\Box\phi$ and $\Diamond\phi$ will also be necessary and thus generally valid.

Consider first a formula such as $\Box\phi$. If $\Box\phi$ is real-world valid, it means that $\Box\phi$ is true in any actual world of every interpretation I. In every world of I, $A\Box\phi$ will be therefore true (in every world, it will be true that in the actual world $\Box\phi$ is true). But then $A\Box\phi$ will be necessary and therefore generally valid. Given the necessary equivalence between $A\Box\phi$ and $\Box\phi$, if $A\Box\phi$ is generally valid, then $\Box\phi$ will also be generally valid.[8]

Consider now a formula such as $\Diamond\phi$. If $\Diamond\phi$ is a real-world valid A-formula, then $\Diamond\phi$ will be true at every actual world of every interpretation. From this, it follows that $A\Diamond\phi$ will be real-world valid, necessary and thus generally valid. But $\Box A\Diamond\phi$ is necessarily equivalent to $\Box A\neg\Box\neg\phi$ which, due to the commutativity of $\neg$ and A, is necessarily equivalent to $\Box\neg A\Box\neg\phi$, which is necessarily equivalent to $\Box\neg\Box\neg\phi$, which is none else than $\Box\Diamond\phi$.

From this, it follows that an A-formula that is real-world valid without being generally valid can only be a non-modal A-formula.

Assume that we only have $\neg$ and $\lor$ as primitive non-modal operators. We can immediately exclude formulas of the form $\neg A\phi$, because such a formula is necessarily equivalent to $A\neg\phi$ and we know already that formulas beginning with A are generally valid, if real-world valid.

We can thus conclude that real-world but not generally valid A-formulas can only be of the form $A\phi \lor \psi$, $\neg A\phi \lor \psi$ or $A\phi \lor \neg\psi$. Where $\psi = \phi$, the first is logically false, while the other are none else than (3). What we can

---

[8]On the equivalence between $A\Box\phi$ and $\Box\phi$ see (Williamson, 1998).

conclude, therefore, is that a formula such as (3) is the *only* case of a real-world, but not generally valid formula for a propositional language such as L^A.

4 On the dispensability of the actuality operator

The actuality operator was introduced in modal languages to satisfy an expressive problem, that of operating (in the propositional case) or quantifying (in the predicative case) outside the scope of modal operators. As we have seen, the problem was solved by introducing a full-fledged, scope-bearing operator, on a par with $\Box$ and $\Diamond$, whose semantics has been defined by the clause (2). The role of A is that of allowing the formula in its scope to be evaluated with respect to the actual world of the interpretation, even in those cases where the clause governed by A is in the scope of $\Box$ and $\Diamond$. The typical behaviour of A is captured by a formula such as:

$$\Diamond(A\phi \land \psi) \qquad (4)$$

where the role of A is that "protecting" ϕ from being evaluated with respect to the world selected by $\Diamond$. But a formula such as (4) could be taken as a circumvoluted way of saying that ϕ and possibly ψ are the case. What (4) expresses thus could be equally captured by a formula such as:

$$\phi \land \Diamond\psi \qquad (5)$$

where no occurrence of A appears. The role of A in (4) is just that signalling where the scope of $\Diamond$ needs to be broken. In this sense, A seems to be more a metalinguistic indicator on how to interpret a formula than a full-fledged component of the formula.

If the contribution of A to a modal language is that of being a device for breaking the scope of modal operators, then one would expect that the role of A be completely superfluous in those formulas where there are no modal operators. The occurrence of A in (3) should thus be taken as superfluous. We have seen, however, that the (extensional) distinction between real-world and general validity is entirely based on the existence of this formula.

Admittedly, the situation is a little bit ironic. By means of languages enriched with A (semantically defined by (2)), we are able to mark the distinction between real-world and general validity; the way in which A helps to mark the distinction, however, is by means of a formula where the role of A seems to be superfluous.

The prime suspect of all this is (2). A natural reaction would be that of taking a formula such as (3) simply as an undesired by-product of an implausible way of giving the semantics of A.[9]

The reason why someone could find implausible a clause like (2) might be explained by means of the distinction, made in the Section 2. A semantic clause such as (2) makes sense only if one has a *sub specie actualitatis* conception of modal truth. Only if one believes that, when we go up in the space of possibilities, plain truth is actual truth (truth in @ of an I) might find plausible the idea that there is an operator by means of which one could speak of the actual world, even from other possible worlds. On the contrary, if one believes that, when we go up in the modal space, plain truth is somewhat "diluted" into possible truth (truth in a w of an I) and that all possible truths are semantically on a par—in the sense that all contribute equally to what is true in an interpretation—then what is eventually plausible is the idea of having an operator by means of which one could speak of *a* world (the one that is contextually relevant), even from another possible world.

In the case of iterated modalities, for example, in the case of a formula such as:

$$\Diamond(\phi \to \Diamond A\psi)$$

the "actualist" A will interpret ψ with respect to w^*, while the "possibilist" A will interpret ψ with respect to the world selected by $\Diamond$. Note that the behaviour of a "possibilist" A is perfectly consonant with the expressive need that A was meant to satisfy (i.e., breaking the scope of modal operators). Though able to satisfy the basic need, the "actualist" A does so at the cost of what could be characterized as "semantic overreaction".

If propositional formulas containing the actuality operator are simply lazy ways of expressing contents that would otherwise be expressible without it, then one should be able to prove that the actuality operator is dispensable. In effect, some dispensability results are available, even if their philosophical significance and applications only very recently have been appreciated.[10] My aim in these final pages is to see whether these dispensability results can be of any use in the real-world/general validity issue.

---

[9]Those who do not accept (2) typically do not accept (3) as a theorem. See, for example, (Gregory, 2001).

[10]The first result of this kind is given in a proof-theoretic manner by Hazen (1978) (even if some traces were already present in (Crossley & Humberstone, 1977)) For another proof-theoretic result see (Stephanou, 2001). A model-theoretic version of the first result is given in

The dispensability results given in (Hazen et al., 2013) proves that every formula of a propositional language L^A is real-world equivalent to a formula without any occurrence of the actuality operator. Two formulas ϕ and χ are real-world equivalent iff $\phi \leftrightarrow \chi$ is real-world valid. This result is proved by proving first that every formula of L^A is strictly equivalent to a combination of so-called A-atoms, where ϕ is strictly equivalent to χ iff $\phi \leftrightarrow \chi$ is generally valid and A-atoms are formulas such as p, Ap, $\Box\phi$ (where ϕ is free of A) and A$\Box\phi$ (where ϕ is free of A). The nice feature of A-atoms is that they are real-world equivalent to formulas without A, such as p or to $\Box\phi$.

As far as the real-world/general validity issue is concerned, what interests us here is just to see onto what A-free formula (3) gets mapped to and whether such a formula is also generally valid. If the real-world valid formula to which (3) is associated is also generally valid, then we can conclude that the real-world/general validity distinction cannot be based on the existence of a formula such as (3).

A formula such as (3) is a conjunction of the form:

$$(\neg A\phi \vee \phi) \wedge (\neg\phi \vee A\phi) \tag{6}$$

Given that we know that A commutes with negation, we know that (6) is (generally and thus real-world) equivalent to:

$$(A\neg\phi \vee \phi) \wedge (\neg\phi \vee A\phi) \tag{7}$$

We know that, by (2), A$\neg\phi$ is real-world equivalent to $\neg\phi$ and that Aϕ is real-world equivalent to ϕ. A formula such as (7) is therefore real-world equivalent to a formula such as:

$$(\neg\phi \vee \phi) \wedge (\neg\phi \vee \phi) \tag{8}$$

But a formula such as (8) is also generally valid. (3) is thus real-world equivalent to a generally valid formula. The occurrence of A in (3) could be dispensed with in favour of a formula like (8). Given that (8) is also generally valid, the distinction between real-world and general validity, as based on (3), seems therefore to be laying on quite a feeble ground.

(Hazen, Rin, & Wehmeier, 2013). A philosophical application of this dispensability result (and a tentative extension to the predicative case) is given by Meyer (2013).

5 Conclusion

In this paper, I have defended the claim that the distinction between real-world and general validity corresponds to a genuine distinction between two conceptions of modal truth, but that the manifestation of this distinction in a modal propositional language like L^A should not to be taken seriously. The reason is that the difference appears, in modal propositional languages, only for formulas with the actuality operator A and the actuality operator is dispensable from such languages.

As shown by Hazen (1976), the dispensability results for A in the propositional case contrast with its apparent non-eliminability in predicate modal logic. This result, however, could be viewed more as a proof of the expressive limitations of first-order modal languages than as a proof of the essentiality of A for such languages.[11]

Should then the distinction between the two notions of modal validity be approached with a pluralistic attitude? Should the difference be conceived as the result of an equivocation of the word 'valid' in the metalogical vocabulary?[12] My claim that a formula like (3) should not be taken seriously as far as the distinction is concerned, could be taken as a way of endorsing this form of logical pluralism, but it is not so. As I have claimed, the distinction seems to be grounded on two substantive views about the nature of modal truth. Choosing one notion of validity or the other, I surmise, is to choose between two substantive conceptions of modality. The real-world/general validity issue could thus be seen as another case where, as T. Williamson writes, "the contentiousness of logic is radical enough to reach the metalogic". (Williamson, 2013, p. 20)

References

Beall, J. C., & Restall, G. (2006). *Logical Pluralism*. Oxford: Oxford University Press.

Chellas, B. (1980). *Modal Logic: An Introduction*. Cambridge: Cambridge University Press.

---

[11] There is no space to pursue the matter here. Notice, however, that, where some of the expressive limitations of first-order languages are overcome, the actuality operator A seems to be dispensable; see (Meyer, 2013).

[12] Meta-logical pluralism seems in effect to be the most recent and appealing form of logical pluralism today; see (Beall & Restall, 2006).

Crossley, J., & Humberstone, L. (1977). The Logic of "Actually". *Reports on Mathematical Logic*, *8*, 11–29.
Davies, M., & Humberstone, L. (1980). Two Notions of Necessity. *Philosophical Studies*, *38*, 1–30.
Gregory, D. (2001). Completeness and Decidability Results for Some Propositional Modal Logics Containing 'Actually' Operators. *Journal of Philosophical Logic*, *30*, 57–78.
Hazen, A. (1976). Expressive Completeness in Modal Logic. *Journal of Philosophical Logic*, *5*, 25–46.
Hazen, A. (1978). The Eliminability of the Actuality Operator in Propositional Modal Logic. *Notre Dame Journal of Formal Logic*, *19*, 617–622.
Hazen, A., Rin, B. F., & Wehmeier, K. F. (2013). Actuality in Propositional Modal Logic. *Studia Logica*, *101*, 487–503.
Hughes, G., & Cresswell, M. J. (1996). *A New Introduction to Modal Logic*. London: Routledge.
Kripke, S. A. (1963a). Semantical Analysis of Modal Logic I: Normal Propositional Calculi. *Zeitschrift für Mathematische Logik und Grundlagen der Mathematik*, *9*, 67–96.
Kripke, S. A. (1963b). Semantical Considerations on Modal Logic. *Acta Philosophica Fennica*, *16*, 83–94.
Menzel, C. (1990). Actualism, Ontological Commitment and Possible World Semantics. *Synthese*, *85*, 355–389.
Meyer, U. (2013). Counterpart Theory and Actuality Operator. *Mind*, *122*, 27–42.
Morato, V. (2014). Actuality and Validity. *Logique & Analyse, forthcoming*.
Nelson, M., & Zalta, E. (2012). In Defense of Contingent Logical Truths. *Philosophical Studies*, *157*, 153–162.
Stephanou, Y. (2001). Indexed Actuality. *Journal of Philosophical Logic*, *30*, 355–393.
Williamson, T. (1998). Iterated Attitudes. *Proceedings of the British Academy*, *95*, 95–133.
Williamson, T. (2013). Logic, Metalogic and Neutrality. *Erkenntnis*, *78*, 257–275.
Zalta, E. (1988, February). Logical and Analytic Truths That Are Not Necessary. *Journal of Philosophy*, *85*, 57–74.

Vittorio Morato
University of Padua
Italy
E-mail: vittorio.morato@unipd.it

Logic and Reasoning

Jaroslav Peregrin[1]

Abstract: Logic, it is often held, is primarily concerned with reasoning; and the conviction that logic and reasoning are two sides of the same coin nowadays usually equates with the conviction that logic spells out some directives for the "right" management of beliefs. In this paper I put forward an alternative view, based on seeing rules of logic as *constitutive* rules, not instructing us *how* to reason, but rather providing us with certain vehicles *in terms of which* to reason. This also emphasizes the social nature of beliefs: they are entities forged in a *social* mold, formed by rules originating from social argumentative practices. Because of this fact, I suggest that trying to understand logic by means of studying (rules of) the kinematics of beliefs of a solitary individual is essentially misguided.

Keywords: logic, reasoning, belief, proof, argument

1 Logic and "belief management"

Logic, it is often held, is primarily concerned with reasoning, and sometimes even defined as *the* science of reasoning. Perkins (2002, p. 187) summarizes the traditional view of logic, nicely, as follows:

> For over two millennia, since the days of Aristotle and Euclid, the notion of formal logic has figured centrally in conceptions of human reasoning, rationality and adaptiveness. To be adaptive, the story goes, we must be rational about ends and means, truth and evidence. To be rational, we must reason about what means suit what ends, what evidence supports what conclusions. And to reason we must respect the canons of logic. It's common to note that transient moments of everyday cognition involve logical moves, at least implicitly. When you hear a dog bark outside, what you hear is a sound you recognize as a bark. You infer the presence of a dog, a deduction that might go:
>
> Around here, only dogs make the sound of a bark.

---

[1] Work on this paper was supported by the grant No. 13-21076S of the Czech Science Foundation.

I hear something that makes the sound of a bark.

Therefore I hear a dog.

Examples like these intimate that formal logic is far more than a playground and workshop for philosophers, mathematicians and designers of microchips. It is, if not the warp and woof of human reasoning, at least the warp, the woof perhaps being the beliefs from which we reason. Insofar as we are successful as a species in ways beyond the reach of chimpanzees, our logical prowess may be the cause.

The common conviction that logic and reasoning are two sides of the same coin nowadays usually equates with the conviction that logic spells out some directives for the "right" management of our system of beliefs, viz. rules that help us weave our web of beliefs in a "correct" way, especially when it is "correct" to incorporate a new belief. (Now of course an all-important question is what exactly the words "right" and "correct" here amount to—some, like Perkins, would try to explain these epithets in terms of a kind of practical success, but most, including Boghossian, are more likely to help themselves to the concept of truth, the nature of which then elicits further questions. However, for us in this paper, such questions are less pressing than the prior question as to whether the laws of logic can be seen as directives of a management of our webs of beliefs at all.)

If the laws of logic are indeed such directives, this would deliver a straightforward answer to the traditional question of the normativity of logic: it is normative in as much as it tells us what it is correct for us to do (for efficiency in coping with our environment, or expanding our collection of true beliefs etc.). Also, it would yield a clear-cut answer to queries concerning the relationship between logic and rationality. If to be rational is to hold only certain patterns of beliefs, which nowadays appears to be the standard view (Way, 2010), then logic is here to tell us how to achieve this. Hence this construal of logical laws appears to put many things into place in one sweep. What more could we want?

There is, however, a major problem with this view, which, I am convinced, renders it ultimately untenable: simply put, there is no sound way of explaining the laws of logic as governing the kinematics of the web of beliefs of an individual. In my opinion, belief is *primarily* a *social* matter, and only secondarily a *personal* one—not in the sense that one cannot believe *privatim*, but because propositions, which constitute the vehicles of

Logic and Reasoning

beliefs, are forged exclusively in a social mold. And I am going to argue that the laws of logic relate more to this mold than to any strategies of dealing with its products within an individual. (To be more precise, I think that *derivatively* these rules may *also* influence the individual belief management; however, this is only a by-product of their primary role.) Hence, though the fact that logic is a theory of reasoning is undoubtedly *in some sense* true, I am convinced that the sense in which it is is actually far more complex than generally assumed, or than the passages quoted above would suggest.

2 Do the rules of logic tell us how to reason?

In a book published about a quarter of a century ago, Harman (1986) insists that logic has very little to do with reasoning, in the sense of "reasoned change of view". For those holding that logic is the science of reasoning, this may be a rather perplexing claim. But on the other hand, if we consider the arguments of Harman, and indeed if we consult obvious facts at hand, we may well start to wonder why we ever thought that logic and reasoning are connected in this straightforward way. Let me quote a more recent paper due to Harman and Kulkarni (2006):

> In the traditional view, reasoning can be modeled by a formal argument. You start by accepting certain premises, you then accept intermediate conclusions that follow from the premises or earlier intermediate conclusions in accordance with certain rules of inference. You end by accepting new conclusions that you have inferred directly or indirectly from your original premises. One problem with the traditional picture is its implication that reasoning is always a matter of inferring new things from what you start out believing. On the contrary, reasoning often involves abandoning things you start out believing. ... You regularly modify previous opinions in the light of new information.

Look at one of the most basic rules of propositional logic, *modus ponens*:

$$A, A \to B \vdash B$$

We can easily imagine somebody using this step in reasoning. She forms a belief, say, that *if it is raining, Tom will take his umbrella when he goes out*, and later she finds *that it is raining*. Hence, using modus ponens, she concludes *that Tom will take his umbrella when he goes out*.

But how exactly is *modus ponens* relevant for our reasoning? That once we have the beliefs A and $A \to B$, we should acquire the belief B? One of the problems pointed out by Harman is that we do not always use *modus ponens* in this way. Imagine that I go home and believe *that my wife is there*. Also I believe that *if my wife is at home, the door is not locked*. I come to the door and find it locked. What I naturally do is give up my belief *that my wife is at home*.

The second, more worrisome problem tabled by Harman is that were we to work out everything that follows from our beliefs, we would never find a place to stop, whereas we obviously only work out what follows when we expect to get something useful.[2] Hence perhaps *modus ponens* does not tell us how we *should* amend our beliefs, but merely how we *may* do so? But if this is so, the rule would tell us how to reason merely in a very indirect, and not very helpful sense. Clearly the space of moves that are in accordance with the laws of logic is abundant; hence we would immediately need some other rules to tell us how to really steer through it. This would seem to compromise seeing the rules of logic as helpful directives for belief management.[3]

MacFarlane (n.d.) considers a third possibility of the normative reading of laws of logic, namely the reading according to which a rule tells us that believing the premises gives us reason to believe the respective conclusion. Also, he distinguishes the scope of the deontic operator: in the case of *ought*, for example, we might read a law of logic so that believing its premises we ought to believe its conclusion; or so that if we ought to believe the premises, we ought to believe the conclusion; or, finally, so that we "ought to see it that" if we believe the premises, we also believe the conclusion. Then he makes one more distinction that cuts through the previous two: this distinction concerns reading the conclusion of the logical law in question

---

[2] There is a further problem (though we will not tackle it in the present paper) with the construal of laws of logic as directives of beliefs management. What we do or do not believe is not generally a matter of our decision; and this would seem to compromise the possibility of seeing it as something that might be reasonably *prescribed* to us by a *rule*. The point is that a rule cannot bind us to do something that we cannot do; hence if we are not capable of changing our beliefs at will, there could hardly be rules that would bind us to do so.

[3] Moreover, we can ask what kind of inappropriateness or sanction would we be liable to were we to disobey *modus ponens* in our mind. Surely not a social one, such as compromising our status as a rational being in the eyes of others, for nobody would know. Hence, would the sanction consist in not being successful in our reasoning? But it is clear that we can imagine circumstances when reasoning *may* be successful even if it ignores any kinds of canons.

either as prescribing us to believe it, or not to disbelieve it. (This gives him, altogether, eighteen possible normative readings of the logical laws.) MacFarlane is inclined to go for "some combination" of (a) *you ought to see to it that if you believe the premises, you do not disbelieve the conclusion* and (b) *you have reason to see to it that if you believe the premises, you believe the conclusion*.

Now, though I think MacFarlane may have isolated the best of the spectrum of options, whether these are acceptable remains at least dubious.[4] While (b) does not seem to quite avoid Harman's objection that it would get our mind clogged with inferences, (a) contains an unclear notion of *disbelieving*. It seems that were we to interpret *disbelief* as simply lack of belief, then *not disbelieving* would simply collapse into *believing* (thus making this first option fall prey to the same problem as the second one), whereas were we to see it more as *believing the opposite*, it would again not tell us anything helpful w.r.t. what to really believe.

Given all of this, it may be good to return to Harman's argument against a straightforward linking of the laws of logic to reasoning. Field (2009, pp. 252–253) summarizes the outcomes of the argument in four points:

1. Reasoning (change of view) doesn't follow the pattern of logical consequence. When one has beliefs $A_1, \ldots, A_n$, and realizes that they together entail B, sometimes the best thing to do isn't to believe B but to drop one of the beliefs $A_1, \ldots, A_n$.

2. We shouldn't clutter up our minds with irrelevancies, but we'd have to if whenever we believed A and recognized that B was a consequence of it we believed B.

3. It is sometimes rational to have beliefs even while knowing they are jointly inconsistent, if one doesn't know how the inconsistency should be avoided.

4. No one can recognize all the consequences of his or her beliefs. Because of this, it is not reasonable to demand that one's beliefs be closed under consequence. For similar reasons, one can't always recognize inconsistencies in

---

[4]It is fair to stress that as MacFarlane's paper remains unpublished (and on his home page, from where it is available, its author points out that he intends to rework it), this cannot be taken as a criticism. In fact it may be even inappropriate to refer to this kind of paper at all; however, the truth is that the systematicity with which this paper sorts out the possible deontic readings of logical laws w.r.t. reasoning is unmatched.

one's beliefs, so even putting aside point 3 it is not reasonable to demand that one's beliefs be consistent.

In view of the problems with logic considered as a theory of "reasoned change of view" we may relish a wholly different interpretation of the laws of logic. Thus Field considers the possibility of seeing the laws of logic as spelling out those forms of inference that necessarily preserve truth (Field attributes this view to Harman; Harman, however, disowns it[5]); but ultimately he dismisses this possibility as unviable. Hence, in the end, Field returns to anchoring the laws of logic in reasoning and ends up with the following probabilistic interpretation (Field, 2009, p. 262; where $P(X)$ denotes the probability of X):

> Employing a logic L involves it being one's practice that when simple inferences $A_1, \ldots, A_n \vdash B$ licensed by the logic are brought to one's attention, one will normally impose the constraint that $P(B)$ is to be at least $P(A_1) + \ldots + P(A_n) - (n-1)$.

But again, this does not seem quite satisfactory. Unlike simply ascribing a belief *simpliciter*, ascribing its probabilistic version (i.e. a probability the ascribee associates with a belief) is a much more complicated matter with much less clear content. Moreover, it is not clear how this avoids the "cluttering up our minds with irrelevancies"; for it seems that it again concerns all the consequences of our beliefs; and adjusting all the relevant probabilities, or even checking them, would again be an infinite process.

3 The social and normative nature of belief

Harman's objections might be understood as protesting merely against the 'direction' in which reasoning works and its 'compulsiveness': it does not (always) proceed from what the usual laws of logic give as their premises to what they give as conclusions. Sometimes it proceeds the other way around (abandoning one of the premises rather than accepting the conclusions); and sometimes it might even proceed leaning upon the laws of logic in more complicated ways. This fact is, I think, important to note (and Harman presents very persuasive arguments against the simplistic construal of reasoning), but what I am going to argue for here is that the problem with the usual view of reasoning lays still somewhat deeper. In particular I think that

---
[5] See (Harman, 2009).

the problem is already when we take for granted that the task of logic is to help us reasonably manage our beliefs, which we have and which it is not the business of logic to explain.

Indeed I think that once we take ready-made beliefs as an unquestioned point of departure of the application of logic, as something that must be explained by something that has nothing to do with logic (perhaps cognitive science?), we are well on the way into a blind alley. What I think is that it is essentially wrong to see logic as a theory of an individual's epistemic achievements. Though, of course, it is an individual who reasons, the ability of reasoning has, I am convinced, an essential social dimension; and logic should be seen as related to this dimension. Recently there has been much attention paid to the question *Is belief normative?*[6] and it seems to me that the answer to this question also gives us the key to understanding the role of the rules of logic w.r.t. reasoning; in particular I believe that what we need is a proper understanding of the sense in which any belief *is* a *normative entity*.

Let me stress, immediately, that I do not see the normativity of belief as stemming from its interconnection with truth (Boghossian, 2003; Weiner, 2005) or knowledge (DeRose, 2002; Williamson, 2000). Instead, I see beliefs as normative in a rather different sense, roughly in the sense in which chess pieces are. Chess pieces are normative entities because they owe their existence (*qua* such) to the rules of chess: it is these rules that make a material vehicle, a piece of wood or ivory or whatever, into a pawn, a rook, or a bishop. And as I see it, it is certain rules, and the rules of logic among them, that make certain material vehicles into 'embodiments of beliefs'.

What are the material vehicles of beliefs? It is often taken for granted that as beliefs are *mental* entities, then if we accept talk about its *vehicles*, then the vehicles must be some items within the brain, perhaps some constellations of electric potentials. However, I think that the direct material vehicles of beliefs are sentences, i.e. certain types of sounds (and/or scribbles). (Of course, this does not contradict the claim that some implementation of beliefs must be in the brain; sentences are also entities that exist somehow via human brains.) And just as it is the rules of chess that make pieces of wood into the chess pieces, so it is the rules of language that make types of sounds into meaningful sentences.

Of course, this raises some questions. Do I mean that there is no belief

---

[6]See, e.g., (Boghossian, 2003; Engel, 2007; Glüer & Wikforss, 2009; Peregrin, 2012; Steglich-Petersen, 2006).

without language? Well, I do; but with two important provisos. First, I mean to say that there is no belief *in our human sense* without language; I do not mean to deny that even language-less brutes may be in states which we may tend to characterize as states of believing something. However, I think that saying about somebody that she believes something in our human sense of the word, involves saying that she knows the place of the belief within the network of many other beliefs (knows, for example, what follows from it or what must be the case for this belief to become true), and language, I am convinced, is the only substratum nourishing enough to sustain such a network.

The other proviso is that though there is no belief without a language, this does not mean that belief would generally be something like an inner assertion. There are, I am convinced, no believers who are not language users; but not every episode of belief must be a matter of language. I understand the constitutive connection between language and belief in the sense of Sellars' "verbal behaviorism": "According to VB [verbal behaviorism]", as he puts it in his characteristically cryptic way (Sellars, 1974, p. 419), "thinking 'that-p', where this means 'having the thought occur to one that-p', has as its primary sense [an event of] saying 'p'; and a secondary sense in which it stands for a short term proximate propensity [dispositional] to say 'p'".[7]

An important consequence is that if all of this is correct, then the rules of logic cannot be seen as *tactical* rules dictating feasible strategies of a game; they are the rules *constitutive* of the game as such.[8] This is a crucial point, because it is often taken for granted that the rules of logic tell us *how to reason* precisely in the tactical sense of the word. But what I maintain is that this is wrong, the rules do not tell us *how* to reason, they provide us with things *with which*, or in *terms of which*, to reason.[9]

---

[7] The parenthetic comments are added by Rosenberg (2009). His exposition of Sellars' view may be also consulted for a more detailed elucidation.

[8] Let me stress that speaking of "constitutive" rules I do not mean *constitutive* as opposed to *regulative* in the terminology of Rawls (1955) or Searle (1969). "Constitutive" rules in the sense entertained here are opposed to "tactical" rules – constitutive rules are those that delimit the space of the game (and it is not important whether they delimit it purely conventionally or on some natural foundations, hence even regulative rules can be seen as constitutive in this sense), whereas tactical rules are those which advice how to move within the space with success.

[9] Another important point, which, however, we will not discuss in this paper, is the dimension of the rejection of the individualist construal of belief (and for that matter, knowledge) yielding the possibility of there being two completely identical individuals such that it would be justified to say of one of them, but not of the other, that she believes something. The point is that somebody's believing something, under this construal, depends not only on his state of mind, but also, as it were, on the social context and especially on what we may call the *in-*

Note that we can think about chess in a similar way. There are two kinds of rules of chess: First, there are rules of the kind that a bishop can move only diagonally and that the king and a rook can castle only when neither of the pieces have previously been moved. These are the rules constitutive of chess; were we not to follow them, then we would not be playing chess. In contrast to these, there are *tactical* rules telling us what to do to increase our chance of winning, rules advising us, e.g. not to exchange a rook for a bishop or to embattle the king by castling. Were we not to follow them, we would still be playing chess, but with little likelihood of winning.

We can imagine the rules of chess as something that produces the pieces, equips them each with its peculiar *modus operandi*, and then see the relevant tactical rules as consisting in setting the individual *modi* into the most efficient teamwork. The rules of logic, viewed analogously, would then have a slightly more complex role: along with furnishing us with logical concepts (each with its peculiar *modus operandi*) they also provide us with a mold in which we cast all other concepts so that they acquire their characteristic shape (and thus can combine with logical ones).[10] Then we face the problem of setting the individual concepts (logical and extralogical) into effective thinking (and we might consider articulating some directives or rules that could then be seen as the tactical rules of reasoning).

To say that the rules of logic are rules not for operating with propositions and beliefs, but rather with sentences (helping make them 'propositionally contentful') is to say that our language games pre-date our reasoning, and indeed pre-date meanings. This might seem a perplexing claim. How could we start to play a language game with *meaningless* sounds? Is not the very point of the games some kind of trafficking of *meanings*? But it only follows that there must have been a certain kind of bootstrapping going on: first proto-games, in which sounds had merely some rudimentary and simple functions (warning, attention attraction) and then some stepwise enhancement of their functionings going hand in hand with the games growing more complex, until they reached their current complicated form with

---

stitutional framework. (Compare believing thus construed with christening. It is obvious that there might be two completely identical individuals, doing completely the same movements, such that it would be justified to say of one of them, but not of the other, that she christens a newborn baby. There is nothing puzzling about the fact that christening can take place only within a clearly delimited institutional framework; and my point is that believing and knowing is, despite appearances, not quite unlike this.) I think that this follows from the fact that our knowledge claims are underlain by what Williams (2001) calls *default and challenge structure*.

[10]The other concepts are not produced by the rules of logic alone, they are co-produced by other kinds of rules. See (Peregrin, 2001) for details.

logical rules and with the functions of words and sentences portrayable as 'expressing concepts' and 'propositions'.

John Searle, reflecting on the origins of language, wrote:

> ...imagine a class of beings who were capable of having intentional states like belief, desire and intention but who did not have a language. What more would they require in order to be able to perform linguistic acts? Notice that there is nothing fanciful in the supposition of beings in such a state, since as far as we know the human species once was in that state. (Searle, 1979, pp. 193–194)

What I have just said involves the exact contradictory of the last claim, namely I see no reason to believe that the human species *ever* was in such a state. Instead of assuming that argumentation is an externalization of reasoning, I am assuming that a certain, relatively recent upgrade of our reasoning faculties is effected by an internalization of argumentation. Thus, in contradistinction to the view of Searle, I concur with Mercier (2010) claiming that "reasoning evolved not to complement individual cognition but as an argumentative device".

4 From proving to reasoning

Let me summarize my view of the matter into the following four points:

1. *Logical rules, and inferential rules in general, are best seen as primarily concerned not with reasoning in the sense of belief management, but with demonstrations and proofs.*

2. *The rules which govern demonstrations and proofs can in turn be seen as rules of certain language games, especially the games which have to do with "giving and asking for reasons".*

3. *Neither demonstration, nor argumentation is an externalization of reasoning (but it can, to a certain extent, be internalized to constitute an extraordinary overlay of our normal reasoning proceedings).*

4. *Hence logical rules are rooted in the regulation of argumentation; the rules are constitutive of the very space of argumentation and consequently of beliefs as inner correlates of assertions – they are constitutive rather than tactical rules.*

The first two points do not seem to be too controversial. As to the first one, when we look at the writings of the most reflective of the founding fathers of modern logic, Gottlob Frege, we can see, from the beginning, that he aims his logical system at proofs, i.e. demonstrations; and he takes pains to stress that this has very little to do with actual reasoning in the sense of what happens in an individual mind. We can see that Frege (1879) saw the laws of logic as laws of reasoning exclusively in the sense of "way of carrying out a proof", which, unlike reasoning in the individualist sense, inevitably has to be public (the question how we can prove something is "more definite" than to be answerable "differently for different persons"). What the laws of logic capture is "not the psychological genesis but the best method of proof".

What is also crucial is that Frege's insistence on keeping with the elementary logical rules when composing proofs is *not* a tactical advise that should help us compose proofs effectively or skillfully. It is an advise that should help us not to leave the realm of logic in the first place; and that should help us make apparent that we are not leaving this realm.

As for the second point, the rules of the usual logical systems, which equip these systems with spaces of possible demonstrations and proofs, can be reframed in game-theoretical terms. Lorenzen (1955) was probably the first to try to erect logic on purely game-theoretical foundations; and the approach has gained in popularity during the beginning of the present century.[11] By fine-tuning the rules of the Lorenzenian games, we can make the games equivalent to various logical systems in the sense that there is a winning strategy for a game associated with a formula just in the case the formula is a theorem/tautology of the corresponding system. Such kinds of games can thus be seen as straightforward implementations of the corresponding logics—or, perhaps more appropriately, the logical systems can be seen as capturing the structure of the corresponding games.

The shift from demonstrations and logical systems to games makes it easier to explain how logic could have come into being. It is plausible that first there were rudimentary language games, which then, by growing in complexity, acquired something as a logical backbone, thereby entangling their sentences into ever more complicated logical interrelationships (such as consequence and incompatibility), and providing for the roles of logically complex sentences (negation as minimal incompatible, conjunction as inferential infimum etc.) The explicitly logical locutions then came into being as

---

[11] See e.g. (Majer, Pietarinen, & Tulenheimo, 2009).

means of explicitly expressing these implicit logical relationships.[12]

Notice that language games are a matter public through and through (as Wittgenstein pointed out in his famous "beetle in the box" example;[13] anything that is principally accessible to only one of the players cannot be part of the game.) Rules of such a game must be public, and hence cannot involve the belief or knowledge of a player (at least, not unless these are also construed as publicly accessible). Importantly, this will exclude rules of the kind of *one may assert p only if one knows or believes that p*.

5 Reasoning as inner argumentation

Let me now turn my attention to the third point. It would seem that the rules of logic must be, primarily, rules of inner reasoning, of which the outer demonstration or the argumentation must be expressive. Overt steps of an argument appear to have to come into being as mirror images of some covert steps we carry out within our minds—if this were not so, the so-called arguments would be mere empty sequences of sounds or scribbles on paper. However, as I have already indicated, I am convinced that this appearance is misguided—'putting the cart before the horse', as it were. I would hold that covert reasoning as a sequence of those steps which are articulated by the laws of logic is much more probably derived from overt argumentation than the other way around.

I have already mentioned Sellars' *verbal behaviorism* as a plausible theory of how such faculties of mind as this proof-like reasoning (as well as, for that matter, propositional thought in general) derived from public practices. Sellars argues that how we construe what happens in our mind in terms of a kinematics of propositions or beliefs (initially from the first-person, but subsequently also from the third-person perspective) is parasitic on how we come to perceive linguistics behavior as the kinematics of utterances.[14]

Davidson (1991) is even more explicit in this respect:

> Until a base line has been established by communication with someone else, there is no point in saying one's own thoughts or

---

[12]This is the so-called *expressivist* account for logic: the idea is that material inferences are more basic than logical ones and that the logical ones are the means of making the material inferences explicit. (See (Brandom, 2000, Chapter 1)).

[13]See (Wittgenstein, 1953, §293). See also (Peregrin, 2011).

[14]Sellars (1956) presents this stance in his much discussed Myth of Jones. See (deVries & Triplett, 2000) for a detailed exposition.

words have a propositional content. If this is so, then it is clear that knowledge of another mind is essential to all thought and all knowledge.

Aside of these philosophical accounts, there have also appeared, recently, more empirically founded findings pointing in the same direction. Thus, for example, Mercier and Sperber (2011) criticize what they see as "the classical view of reasoning", which has it that the principal function of reasoning is to enhance individual cognition.[15] Mercier and Sperber argue on empirical grounds that this must be rejected because, far from purifying the mind of mistaken beliefs, reasoning itself can bring in new mistakes, and over and above this, it often rationalizes the existing beliefs instead of correcting them.

These reasons are compelling, and justify the authors in putting forward an alternative to the "classical view": what they propose is that the emergence of reasoning is best understood within the framework of the evolution of human communication:

> Reasoning enables people to exchange arguments that, on the whole, make communication more reliable and hence more advantageous. The main function of reasoning, we claim, is *argumentative* ... Reasoning ... enables communicators to produce arguments to convince addressees who would not accept what they say on trust; it enables addressees to evaluate the soundness of these arguments and to accept valuable information that they would be suspicious of otherwise. Thus, thanks to reasoning, human communication is made more reliable and more potent. (Mercier & Sperber, 2011, pp. 60–72)

Thus, when we internalize the laws of argumentative language games we are facilitated to do covertly what was previously overt: namely, to convince an audience by citing reasons. In this way, we gain a specific overlay to our prior reasoning faculties, an overlay to which we take recourse when solving certain specific tasks, or when we want to check meticulously the conclusions achieved by means of ordinary reasoning. This new skill, however, does not displace our original ways of reasoning, nor diminish their

---

[15] Recently, this view has acquired the shape of a theory that reasoning is a matter of an evolutionary younger module of the mind/brain that had developed for the purposes of rehearsing and correcting the mistakes of an evolutionary older, swifter and intuitive system of control of behavior. See, e.g., (Evans & Frankish, 2009).

import—it is something that we do not use very frequently (if for no other reason than that it is time consuming, and most of our reasoning must be done in the 'on-line' mode).

I think that counterintuitive as this view might seem at first sight, it has a lot of plausibility to it. The problem many people would have with it is that while it seems to be a plain fact that it is an *individual* that reasons (in her, as it were, 'foro interno'), and that it is only an individual that can forge *meaningful* sentences to do the reasoning with. The view put forward here does not reject the first point—of course it is an individual that does the reasoning (though un-internalized argumentation done by a group of people can be perhaps also called reasoning). However, it does challenge the second point: meanings are brought into the mind from a public space where they are forged within the furnace of human interaction.

6 Laws of logic as constitutive

The individualistic approaches to logic (the "traditional views" criticized by Mercier and Sperber) take for granted that logic spells out tactical, rather than constitutive rules. Presumably, this is because *prima facie* there are no obvious alternatives to this construal of the rules of logic. Upgrading beliefs does not seem, on the surface, to be a game, at least not a game with any similarity to chess, *viz.* a game *constituted* by rules.

However, let us reappraise how we look at the constitutive rules of chess. We may see them, as we have before, as constituting the *pieces* as such: kings, rooks, bishops etc. Once we have these items, each of them coming with a specific 'behavior' (thus the bishop with the propriety of moving diagonally etc.), we can forget about the constitutive rules and see the space of chess as delimited by whatever it is possible to achieve with them. And the achievements are non-trivial, though they are usually not particularly important for us, chess not generally playing a significant role in our lives. Now the idea is that our beliefs are analogous to the pieces; that our tactics for dealing with them are based on the natures of the beliefs, these natures being established by constitutive rules.[16] And here the achievements we can reach when we learn to orchestrate beliefs efficiently are not only important, but also highly non-trivial: they help us steer clear of the perils of our world, and enjoy what it has to offer, much more effectively than before.

---

[16] See (Peregrin, 2010) for a more detailed exposition of this.

This explains why the rules of logic do not really tell us how to reason, at least not in a very nontrivial sense—it is for the same reason that the rules of chess do not tell us how to play chess, except in the trivial sense that they tell us what are the permitted moves. To learn how to play chess we need another kind of rules (or guidances), the tactical ones. The former rules merely set up the stage, or produce the characters with which to play; it is only the latter ones that tell us what to do.

The fact that a proof or a demonstration consists of steps according to these very rules does not mean that this would be what we actually do when we reach new beliefs in our heads; it is a matter of the fact that a demonstration as such must be utterly transparent, in particular it must be clear that all its steps are legitimate. And the best way to make them clearly legitimate is to make them directly accord with the elementary rules. On the other hand, the fact that we have a lot of potential steps sanctioned directly by the rules, does not actively help us to chain such steps together appropriately to get a proof of a given claim. If this chaining were a matter of rules, then they would have to be rules very different from the constitutive rules which we borrow to assemble the proofs from.

We often say that human thought differs from that of other animals (insofar as these can be ascribed something as thought at all) because it is *conceptual*: that we humans, in contrast to our animal cousins, have *reason* (and hence are able *to* reason), that we can think and infer *logically* etc. The picture at which we have arrived here, suggests that logic is a kind of tool enhancing our thought 'from without'—we have developed certain complicated and useful social practices, crucially involving language, and these practices equipped us with certain tools that we later internalized. The tools are logical concepts that help us organize and effectively maintain what we know and what we believe.

7 Conclusion

I have argued that logical rules are not a matter of a strategy of an optimal belief management done by an individual; I have urged that they have an essential social dimension. The dimension does not make the individual dependent on the society in that the conclusions she reaches are not her own, but rather in that the very vehicles of reasoning are originally of a social making. The conjecture put forward and defended in this article is that these rules originated as rules of demonstrations and proofs, hence as rules of certain (argumentative) language games. Inner reasoning, then is the internalization of public argumentation (rather than the other way around) – it is not that every reasoning would be a chain of covert assertions following one from another, but rather that every reasoning takes place on the conceptual and propositional level and hence uses vehicles that originated in public language games. The most important point to which this train of thought has led us is that logical rules, are *constitutive* rules, they are not *tactical* rules for dealing with beliefs and other propositions, but rather rules that are responsible for there being something as propositions in the first place.

References

Boghossian, P. (2003). The Normativity of Content. *Philosophical Issues*, *13*, 31–45.

Brandom, R. (2000). *Articulating Reasons*. Cambridge, MA: Harvard University Press.

Davidson, D. (1991). Three Varieties of Knowledge. In A. Griffiths (Ed.), *A. J. Ayer: Memorial Essays* (pp. 153–166). Cambridge: Cambridge University Press. (Reprinted in Davidson, D. (2001). *Subjective, Intersubjective, Objective*. Oxford: Clarendon Press, 205–220.)

DeRose, K. (2002). Assertion, Knowledge, and Context. *The Philosophical Review*, *111*, 167–203.

deVries, W. A., & Triplett, T. (2000). *Knowledge, Mind and the Given*. Indianapolis: Hackett Publishing Company.

Engel, P. (2007). Belief and Normativity. *Disputatio*, *23*, 179–203.

Evans, J., & Frankish, K. (Eds.). (2009). *In Two Minds: Dual Processes and Beyond*. Oxford: Oxford University Press.

Field, H. (2009). What Is the Normative Role of Logic? *Proceedings of the Aristotelian Society*, *83*, 251–268.

Frege, G. (1879). *Begriffsschrift*. Halle: Nebert. (English translation available in J. van Heijenoort (ed.), *From Frege to Gödel: A Source Book in Mathematical Logic, 1879–1931* (pp. 1–82). Cambridge, MA: Harvard University Press, 1967.)

Glüer, K., & Wikforss, A. (2009). Against Content Normativity. *Mind*, *118*, 31–70.

Harman, G. (1986). *Change in View (Principles of Reasoning)*. Cambridge, MA: MIT Press.

Harman, G. (2009). Field on the Normative Role of Logic. *Proceedings of the Aristotelian Society*, *109*, 333–335.

Harman, G., & Kulkarni, S. R. (2006). The Problem of Induction. *Philosophy and Phenomenological Research, 72*, 559–575.

Lorenzen, P. (1955). *Einführung in die Operative Logik und Mathematik.* Berlin: Springer.

MacFarlane, J. (n.d.). *In What Sense (if Any) Is Logic Normative for Thought?* (draft available from the author's web page)

Majer, O., Pietarinen, A., & Tulenheimo, T. (Eds.). (2009). *Games: Unifying Logic, Language, and Philosophy.* Berlin: Springer.

Mercier, H. (2010). The Social Origins of Folk Epistemology. *Review of Philosophy and Psychology, 1*, 499–514.

Mercier, H., & Sperber, D. (2011). Why Do Humans Reason? (Arguments for an Argumentative Theory). *Behavioral and Brain Sciences, 34*, 57–111.

Peregrin, J. (2001). *Meaning and Structure.* Aldershot: Ashgate Publishing.

Peregrin, J. (2010). Logic and Natural Selection. *Logica Universalis, 4*, 207–223.

Peregrin, J. (2011). The Use-theory of Meaning and the Rules of Our Language Games. In K. Turner (Ed.), *Making Semantics Pragmatic* (pp. 183–204). Bingley: Emerald.

Peregrin, J. (2012). Inferentialism and the Normativity of Meaning. *Philosophia, 40*, 75–97.

Perkins, D. N. (2002). Standard Logic as a Model of Reasoning: The Empirical Critique. In D. M. Gabbay, R. H. Johnson, H. J. Ohlbach, & J. Woods (Eds.), *Handbook of the Logic of Argument and Inference* (pp. 186–223). Amsterdam: Elsevier.

Rawls, J. (1955). Two Concepts of Rules. *Philosophical Review, 64*, 3–32.

Rosenberg, J. F. (2009). Wilfrid Sellars. In E. N. Zalta (Ed.), *The Stanford Encyclopedia of Philosophy* (Fall 2010 ed.).

Searle, J. R. (1969). *Speech Acts: An Essay in the Philosophy of Language.* Cambridge: Cambridge University Press.

Searle, J. R. (1979). Intentionality and the Use of Language. In A. Margalit (Ed.), *Meaning and Use* (pp. 181–197). Dordrecht: Reidel.

Sellars, W. (1956). Empiricism and the Philosophy of Mind. In H. Feigl & M. Scriven (Eds.), *Minnesota Studies in the Philosophy of Science, Volume I: The Foundations of Science and the Concepts of Psychology and Psychoanalysis* (pp. 253–329). Minnesota: University of Minnesota Press.

Sellars, W. (1974). Meaning as Functional Classification. *Synthese, 27*, 417–437.

Steglich-Petersen, A. (2006). The Aim of Belief: No Norm Needed. *The Philosophical Quarterly*, *56*, 500–516.
Way, J. (2010). The Normativity of Rationality. *Philosophy Compass*, *5*, 1057–1068.
Weiner, M. (2005). Must We Know What We Say? *The Philosophical Review*, *114*, 227–251.
Williams, M. (2001). *Problems of Knowledge: A Critical Introduction to Epistemology*. Oxford: Oxford University Press.
Williamson, T. (2000). *Knowledge and Its Limit*. Oxford: Oxford University Press.
Wittgenstein, L. (1953). *Philosophische Untersuchungen*. Oxford: Blackwell.

Jaroslav Peregrin
Institute of Philosophy, Czech Academy of Sciences
The Czech Republic
E-mail: `jarda@peregrin.cz`

Strange Case of Dr. Soundness and Mr. Consistency

MARIO PIAZZA
AND GABRIELE PULCINI[1]

Abstract: We scrutinize the well-known route to the consistency of a formal system *via* its soundness with respect to the intended model or, more generally, to the classical bivalent notion of truth. In particular, we argue that in the case of first-order Peano Arithmetic this route does not succeed because of severe circularities, apart from reducing important epistemological questions to triviality. Since our argument has a general scope, it purports to embarrass the traditional consistency-*via*-soundness proofs provided for *any* formal system whatsoever. Consistency indeed should be regarded as a prior notion with respect to soundness from both the epistemological and logical points of view. Such reflections motivate a different conceptualization of soundness, that is a way of understanding this property from the inside, eschewing any reference to an external structure of "higher-level" unformalized mathematical truths.

Keywords: soundness, consistency, Peano arithmetic, mathematical truth, foundations of mathematics.

1 Preamble

The question of the logico-epistemological interplay between soundness and consistency is a deeply important and perplexing one. Yet mainstream thinking about this interplay seems to be anything but fine-grained in that it ignores the philosophical cargo. It is commonly held that the soundness of a given formal theory T *implies* its consistency. The standard version of the proof runs by contraposition: suppose T inconsistent; then, for a certain statement α, we have that T proves both α and $\neg\alpha$. Now, under the classical bivalent notion of truth, either α or $\neg\alpha$ must be true and so T is unsound insofar as it proves a false statement. On the other hand, it is plain that consistency does not imply soundness as a corollary of Gödel's first incompleteness theorem. In fact, consider first order Peano Arithmetic (PA) and

---

[1] Work partially supported by Autonomous Region of Sardinia under grant L.R.7/2007 CRP-17285 (TRICS). The authors wish to thank Volker Halbach, Jaroslav Peregrin and Jean-Yves Girard for their useful comments on some issues raised in this paper.

let $\mathcal{G}$ be the undecided Gödel sentence. If PA is consistent, then PA $\cup$ $\{\neg\mathcal{G}\}$ is consistent as well. Now, $\mathcal{G}$ is interpreted as a true proposition by the standard model and so PA $\cup$ $\{\neg\mathcal{G}\}$ is consistent but unsound[2].

However, it is exactly at the level of highly significant theories like PA that one must feel uneasy about the transition from soundness to consistency. Michael Dummett qualified the proof of the consistency of PA via its soundness as *not genuinely informative* (Dummett, 1963). Yet this litotetic expression has the ring of a paradox, for the two terms "proof" and "uninformative"—as Crispin Wright (1994) pointed out—stand in clear opposition to each other. How can we make sense of an "uninformative proof" as a mathematical/epistemological object? There seem to be two competing possibilities: we can either identify this sort of argument with a proof under some unverified hypotheses, or with an ill-founded one. Undoubtedly, the first option has had many more sympathizers among logicians and philosophers: an uninformative proof is an "open" proof, namely a proof under a hypothesis in search of validation. And, of course, this is supposed to be the hypothesis that the arithmetical axioms are true. Thus, as soon as the soundness of PA has been endorsed, consistency will come to the fore with the same degree of certainty. On this view, then, the consistency-*via*-soundness proof (CvS-proof, for short) counts as a logically sound argument, though developed in the shadow of some "undischarged" hypothesis, so to speak. As you get the truth of the axioms, the consistency of the entire system will come along as a complementary property.

In this paper, we entertain and defend the second and more radical option: the CvS-proof is uninformative in the sense of being ill-founded. Indeed, we spot some circularities which afflict the conceptual structure of the standard CvS-proof. This amounts to acknowledging that consistency is *logically prior* to the soundness property, or—in the epistemological dimension—that our grasp of the soundness of a given theory *presupposes* our grasp of its consistency. The conclusion to which our analysis points is that the soundness property should be revised along an internal syntactical axis, without any reference to an external structure of "higher-level" unformalized mathematical truths.

---

[2] For a recent restatement of this argument see (Halbach, 2011; Isaacson, 2011).

Soundness & Consistency 163

2 The CvS-proof under focus

2.1 The canonical reading

Let us briefly recall the well-known soundness proof of PA with respect to its standard model $\mathcal{N}$.

Theorem 1 (Soundness) *If* $\text{PA} \vdash \alpha$*, then* $\mathcal{N} \vDash \alpha$.

Proof. The proof proceeds by induction on the length of the proof of α in PA.

- *Basis.* Each axiom of PA formalizes an arithmetical truth, i.e. each axiom is true in $\mathcal{N}$[3].

- *Step.* Inference rules are valid: they preserve truth from the premisses to the conclusion. When deductively expressed *à la* Hilbert, PA contains only the *modus ponens* as inference rule:

$$\frac{\alpha \quad \alpha \to \beta}{\beta}$$

In this way, if α and $\alpha \to \beta$ are both arithmetical truths, then it follows straightforwardly that β too is an arithmetical truth.

□

Now, the consistency proof of PA comes from soundness as a very easy corollary.

Corollary 1 PA *is a consistent theory.*

Proof. Suppose PA is not consistent. Then, there is a statement α such that $\text{PA} \vdash \alpha$ and $\text{PA} \vdash \neg\alpha$. Now, for bivalence, either α or $\neg\alpha$ is a *true* statement (or: either α or $\neg\alpha$ is a *false* statement). Therefore, PA proves a false statement, which contradicts the fact that PA is a sound theory. □

---

[3]For instance, take the following PA axiom:

$$\forall x \forall y (x = y \to \mathsf{S}(x) = \mathsf{S}(y)).$$

When S is interpreted by the standard model as the successor function, the axiom is trivially verified, that is:

$$\mathcal{N} \vDash x = y \to (x+1) = (y+1).$$

Concerning the soundness proof, Jean-Yves Girard complains of epistemological emptiness:

> If A is provable in PA, then A is *true*; this is not the result of deep analysis of arithmetic, but simply the remark that the axioms are true, and that the rules of predicate calculus preserve truth. From the *epistemological* standpoint, this remark has absolutely no value at all, because we have precisely agreed on axioms and rules that we believe to be true. (And the average mathematician does not even know the distinction between provability and truth.) Hence the fact that theorems of arithmetic are true appears to be a trifle (Girard, 1987, pp. 215-216).

We agree. The rest of this section is devoted to explaining why the whole procedure for achieving the consistency of PA by means of its soundness is not a show of logical (and epistemological) innocence.

2.2 En attendant Gödel

The line of our argument begins with a passage from Gödel's 1951 Gibbs lecture, where Gödel points out the connection between our validation of the axioms of a formal theory like PA and our commitment to their consistency. The Second Incompleteness Theorem

> makes it impossible that someone should set up a certain well-defined system of axioms and rules and consistently make the following assertion about it: "All of these axioms and rules I perceive (with mathematical certitude) to be correct and moreover I believe that they contain all of mathematics." If someone makes such a statement he contradicts himself. *For if he perceives the axioms under consideration to be correct, he also perceives (with the same certainty) that they are consistent* [emphasis added]. Hence he has a mathematical point not derivable from his axioms (Gödel, 1995, p. 309).

For the purposes of our discussion, the looming problem is how to spell out the italicized conditional sentence which carries a note of warning. What Gödel literally says is that the perception of axiomatic truths and that of their consistency behave like entangled but not overlapping cognitive states: the unfolding of the former perception leads to the latter. The above conditional, however, asks to be understood as expressing a sort of epistemic necessity:

Soundness & Consistency 165

our commitment to the consistency of the axioms is the only guide to the perception of their truth. To see why one only has to bear in mind that the perception invoked by Gödel is accompanied by mathematical certainty and raise the following question accordingly: what constrains the genesis of the special perception-*cum*-mathematical certainty that the axioms $A_1, ..., A_n$ are true? The answer is that nothing is more suitable than consistency itself insofar as the above perception necessarily involves that of falsity of $\neg A_1, ..., \neg A_n$.

Gödel's remark sets the stage for arguing that the route to consistency is, so to speak, "consistency-laden" for any epistemic agent taking her belief that the axioms are true seriously. In the face of this sort of route, perhaps an advocate of the CvS-proof might reply that we do not need to have mathematical certainty that the axioms are true *before we prove* consistency and, therefore, that any reference to consistency at the inductive basis of the soundness theorem can be avoided. So, the objector could offer a two-level reading of the CvS-argument, namely a reading similar to the one proposed by Alan Gewirth (1941) for the Cartesian Circle in the Third *Meditation*. The idea is that it is only at the end of the CvS-proof that one achieves the *mathematical* certainty of the truth of the axioms by attaining, at the beginning, *psychological* certainty, in the sense of being psychologically incapable of doubting the axioms.

Unfortunately, the matter is not so simple. First, it is hard to understand how the higher, objective, level of certainty could be reached by starting off with a lower, subjective one, although the onus of explaining this epistemic movement is on the objector. Second, and most important, Gödel's epistemological point can be logically assessed when the structure itself of the CvS-proof is examined closely.

2.3 Consistency and Tarskian truth

Let us now proceed to unpack the very inductive structure of the soundness proof. Indeed, the central feature of this argument is the lack of any specific content, except that of trivially restating the induction principle itself: *if* a certain property $P(x)$ holds for all the axioms *and* permeates each inference rule $\mathcal{R}$, i.e.,

$$\frac{\beta_1 \ \ ... \ \ \beta_n}{\beta} \mathcal{R} \implies \frac{P(\beta_1) \ \ ... \ \ P(\beta_n)}{P(\beta)} \mathcal{R},$$

then P holds for any theorem of PA.

Such an empty inductive shell becomes a soundness proof, and so a consistency proof, once $P(x)$ is supposed to stand for the Tarskian predicate representing the notion of truth-in-$\mathcal{N}$. Now, since all theorems of PA are true, it is clear that the provability of any formula α implies the unprovability of its negation $\neg\alpha$. For the sake of clarity, the critical deductive juncture is expounded below:

1. $\vdash \alpha \Rightarrow \mathcal{N} \vDash \alpha$ *soundness*
2. $\mathcal{N} \vDash \alpha \Rightarrow \mathcal{N} \nvDash \neg\alpha$ *(contraposed) Tarskian truth definition*
3. $\vdash \alpha \Rightarrow \mathcal{N} \nvDash \neg\alpha$ *transitivity 1,2*
4. $\mathcal{N} \nvDash \neg\alpha \Rightarrow \nvdash \neg\alpha$ *(contraposed) soundness*
5. $\vdash \alpha \Rightarrow \nvdash \neg\alpha$ *transitivity 3,4.*

Now, it is easy to acknowledge that the consistency statement finally achieved

$$\vdash \alpha \Rightarrow \nvdash \neg\alpha$$

turns out to be nothing else but the Tarskian truth definition for negations

$$\mathcal{N} \vDash \neg\alpha \Leftrightarrow \mathcal{N} \nvDash \alpha$$

in its contraposed one-way form:

$$\mathcal{N} \vDash \alpha \Rightarrow \mathcal{N} \nvDash \neg\alpha$$

So, we end up with a form of circularity launched by the very fact that the Tarskian truth definition for negations disquotationally implies the consistency itself[4]. If this is right, then this circularity lies in the way of any attempt to prove the consistency of a formal system by appealing to the semantical notion of Tarskian truth. Moreover, to the extent that the truth of A implies the falsity of $\neg A$ the Tarskian definition logically fits Gödel's epistemological point that the perception of consistency is the precondition for that of truth.

---

[4] As stressed in section 2.1, the CvS-proof standardly employs bivalence in order to convert soundness into consistency. In PA the principle of bivalence can be presented through the following two clauses: for any formula α,

1. $\mathcal{N} \vDash \alpha$ or $\mathcal{N} \vDash \neg\alpha$ (either α or its negation is true),
2. $\mathcal{N} \vDash \alpha \Rightarrow \mathcal{N} \nvDash \neg\alpha$ (there are only two truth-values: *true* and *not-true* (= false)).

Put in these terms, the Tarskian biconditional $\mathcal{N} \vDash \alpha \Leftrightarrow \mathcal{N} \nvDash \neg\alpha$ just expresses bivalence: the rightward direction is given by (2), while the leftward one can be easily restated as (1). Now, we can get consistency from soundness just by resorting to (2), without assuming full bivalence.

2.4 Against the "Tarskian diagnosis"

Hartry Field has called "Tarskian diagnosis" the following explanation about the impossibility of achieving a *formalized* CvS-proof (which he terms the "Consistency Argument") within PA:

> If T is a theory like Zermelo-Fraenkel set theory ZF [or, of course, PA], the diagnosis of the breakdown of the Consistency Argument is simple: theories like this do not contain a general truth (or satisfaction) predicate, so the argument cannot even be formulated in the theory (Field, 2006, p. 568).

Nevertheless, we argue that a plausible account for the breakdown demands more than the appeal to the expressive limits of PA as they are demarcated by Tarski's theorem. To see the failure of the Tarskian diagnosis, we then need to take into account PA^{Tr}, i.e. the theory obtained from PA by "internalizing" the Tarskian truth predicate, so as to meet Field's requirement. In a nutshell, PA^{Tr} is obtained from PA by: (i) extending the language by means of the truth predicate $Tr(x)$ and (ii) adding a cluster of axioms *à la* Tarski for deductively managing the new predicate. In PA^{Tr} the formal statement

$$\forall x((Theor_{PA}(x) \wedge Sentence_{PA}(x)) \rightarrow Tr(x))^5$$

representing the soundness of PA is a theorem (Halbach, 2011). This means that PA^{Tr} is both *expressive enough* to formalize the soundness property and *deductively strong enough* to prove it.

Now, the Gödelian consistency statement

$$Cons_{PA} \equiv \neg Theor_{PA}(\ulcorner \bot \urcorner)$$

is also a theorem of PA^{Tr}, insofar as it just comes as a corollary of the formal soundness statement. Indeed, the provability of the formal soundness implies the provability of the so-called *Local Reflection Principle*: for any sentence α:

$$PA^{Tr} \vdash Theor_{PA}(\ulcorner \alpha \urcorner) \rightarrow \alpha$$

from which it is easy to derive the Gödelian sentence $\mathcal{G}$ which is, by the Second Incompleteness Theorem, provably equivalent to $Cons_{PA}$.

---

[5] Here $Theor_{PA}(x)$ and $Sentence_{PA}(x)$ stand for the standard Gödelian provability predicate and the predicate which deductively recognizes the Gödelian coding of a well formed PA formula, respectively.

So far so good. Nevertheless, a question resurfaces. Namely, the question of whether we are entitled to regard this kind of formalized CvS-proof (unlike its unformalized counterpart) as reliable. The answer is a big "no". The reason is that the theorems of a given theory cannot be trusted without the guarantee that the theory at issue does not collapse into triviality, i.e. once a well-established consistency proof is supplied. Indeed, classical inconsistent theories allow any proposition to be proved, losing in this way any kind of informativeness.

Let us summarize our new puzzling situation as follows. On the one hand, the acceptance of the consistency statement $Cons_{PA}$ crucially relies on the trustworthiness of the formal soundness expressed within PA^{Tr}; on the other, one needs to *assume* the consistency of PA^{Tr} in order to trust all its theorems and, specifically, the formal soundness itself. All that needs emphasizing here is that Field's Tarskian diagnosis misconceives the very nature of a phenomenon that has a more general significance. Indeed, the breakdown is not due to the lack of expressive resources, but rather it is endemic to the whole enterprise of the CvS-proofs: one cannot hope to prove the soundness of any formal theory whatsoever without a commitment to its consistency.

3 Full statement of the case

3.1 An alternative diagnosis

As we have urged in the preceding sections, the CvS-proof is an ill-founded mathematical argument *running under the hypothesis of the consistency itself*. Such a circularity manifests itself at two different levels: one epistemological and the other logical.

1. The epistemological level concerns the act of perceiving our list of axioms as self-evident mathematical truths. As stressed by Gödel, such a cognitive achievement needs to be supported by the (more or less implicit) assumption of consistency. Without the belief in the consistency, the perception of the truth cannot be achieved with full mathematical certitude. Thereby, when epistemologically considered, the soundness proof remains a proof under the hypothesis of the consistency of the whole system.

2. The logical level concerns the argument turning soundness into consistency. Indeed, the classical Tarskian notion of truth encapsulates

the notion of consistency itself, so as to disguise it. Therefore, the consistency of the system is presupposed whenever one deals with arguments stressing the Tarskian truth predicate.

In brief, far from being a corollary of the soundness property, consistency is a *necessary* condition for achieving the soundness itself. Please observe that (under the hypothesis of the consistency of PA!) it is not a *sufficient* condition. Actually, a theory like PA ∪ {¬$\mathcal{G}$} turns out to be consistent albeit unsound.

At this stage of the discussion the pervasive misunderstanding about the logical hierarchy between soundness and consistency cries out for explanation. One of the most pervasive ideas from early twentieth-century logic is that the consistency of a formal theory is a property along two dimensions, semantical and syntactical, whereas soundness is exclusively a semantical notion: sound theories are those whose theorems formally stand for mathematical or logical *truths*[6]. Our suspicion is that the above misunderstanding arises from regularly conflating semantical (*sem-*) and syntactical (*syn-*) consistency. Call a theory T "*sem*-consistent", if there is a model $\mathcal{M}$ in which T is realized, and call it "*syn*-consistent" if no contradiction is derivable. Once this distinction is introduced, there is no room for misunderstanding. Therefore, it should now be clear that what we trivially obtain from soundness is *sem*-consistency, insofar as the notion of soundness itself involves a reference model. But it is *syn*-consistency that the familiar CvS-proof attempts to establish. Incidentally, let us stress that *syn*-consistency is stronger than its semantical counterpart. There are standard ways to construe a set-theoretical model for any *syn*-consistent first-order theory, but you need semantical completeness in order to derive *syn*-consistency from *sem*-consistency. By the First Incompleteness Theorem, PA lacks semantical completeness, so that the two different renditions of the consistency property cannot be reconciled.

3.2 Soundness from the inside

Such reflections push us toward a different conceptualization of soundness. In effect, while we have no good reasons for thinking that soundness should be confined to semantics, we have *good* reasons for thinking that it should dissociate from it. So, it seems we cannot escape the question of whether

---

[6]Pressing the historical point, even if relevant to the issues we want to raise, would take us too far afield.

it is possible to characterize the soundness property from the inside, from within a purely proof-theoretical arena, restoring the deductive hierarchy dismantled in the previous sections.

Stripped of its semantical features, the soundness property can be reformulated as follows:

> *A theory* T *is sound if any theorem of* T *can be accepted as a reliable, trustworthy, mathematical (or logical) utterance.*

Put in these terms, our final task is that of characterizing such an informal view through a purely proof-theoretical approach. We propose to *internalize* the notion of soundness by means of the cut-elimination property[7]. Here, the adjective "internal" should be understood as meaning that the system T is sound with respect to T itself, without any reference to an external structure of "higher-level" unformalized mathematical truths. Thus, our definition of internal soundness becomes the following:

> *A formal system* T, *once expressed in a Gentzen-style formalism, is* sound *when the cut-rule is eliminable.*

Cut-elimination usually has two different proof-theoretical sides, one static and the other dynamic. *Analyticity* of proofs constitutes the static side and it refers to the fact that any theorem can be analytically proved by means of a proof enjoying the subformula property: "no concepts enter into the proof other than those contained in its final result", so that "the final result is, as it were, gradually built up from its constituent elements" (Gentzen, 1969, p. 69, p. 88). The dynamical aspect concerns *normalization*, i.e. the specific algorithm effectively able to transform a proof with cut-rules into a proof without them. Both these different aspects of cut-elimination seem to provide a more subtle characterization of the notion of soundness. On the one hand, analyticity allows us to get consistency in a non-trivial and computationally informative way. On the other, normalization guarantees that formal arguments resorting to lemmas can be converted into equivalent arguments without lemmas, so as to make *explicit* their really useful information. In other words, a theorem α of a cut-free theory T is sound w.r.t. T itself insofar as any formal justification of α can be algorithmically explicated. Vice versa, the reliability of the systems without the cut-elimination property is not measurable with respect to the complexity of the associated

---

[7]Such a proposal is consonant with some recent considerations on philosophy of logic developed by Jean-Yves Girard (2011).

normalization procedure, so that their (internal) soundness remains unaddressable.

It is well-known that normalization is an algorithm whose computational complexity partially depends on the specific formal system to which it applies. Therefore, soundness may be now graded in point of the complexity of the corresponding normalization procedure. In this way, the logical notion of soundness, once internalized, ceases to be a static property for becoming a *yardstick* for the reliability of formal systems. For instance, the normalization procedure for PA is much more complex than that for intuitionistic and classical logic, due to the fact that, by the Second Incompleteness Theorem, the normalization algorithm for PA is not even expressible by means of a provably (in PA) total recursive function.

The temptation exists to claim that soundness is a matter of degree. Purely logical theorems should be recognized as *more* sound, i.e. as having an higher reliability degree, than theorems of formal arithmetic. Indeed, classical logic is patently consistent, whereas the consistency of PA still remains a controversial question depending on the mathematical trust you put in Gentzen's proof. The fact is that normalization procedures for PA have an unmanageable complexity and this computational aspect is indeed clearly reflected in the Second Incompleteness Theorem.

References

Dummett, M. (1963). The Philosophical Significance of Gödel's Theorem. *Ratio*, *5*, 140–155. (Reprinted in Dummett, M. (1978). *Truth and Other Enigmas*. London: Duckworth.)

Field, H. (2006). Truth and the Unprovability of Consistency. *Mind*, *115*, 567–605.

Gentzen, G. (1969). *Collected Papers*. Amsterdam: North Holland.

Gewirth, A. (1941). The Cartesian Circle. *The Philosophical Review*, *50*, 368–395.

Girard, J.-Y. (1987). *Proof Theory and Logical Complexity*. Naples: Bibliopolis.

Girard, J.-Y. (2011). *The Blind Spot*. Zürich: Europ. Math. Society.

Gödel, K. (1995). Some Basic Theorems on the Foundations of Mathematics and Their Implications. In S. Feferman, J. W. Dawson, W. Goldfarb, C. Parsons, & R. Solovay (Eds.), *Collected Works vol. III: Unpublished Essays and Lectures* (pp. 304–323). Oxford: Oxford University Press.

Halbach, V. (2011). *Axiomatic Theories of Truth*. Cambridge: Cambridge University Press.
Isaacson, D. (2011). Necessary and Sufficient Conditions for Undecidability of the Gödel Sentence and Its Truth. In D. DeVidi, M. Hallett, & P. Clark (Eds.), *Logic, Mathematics, Philosophy, Vintage Enthusiasms* (pp. 135–152). Dordrecht: Springer.
Wright, C. (1994). About "The Philosophical Significance of Gödel's Theorem": Some Issues. In B. McGuinness & G. Oliveri (Eds.), *The Philosophy of Michael Dummett* (pp. 167–202). Dordrecht: Kluwer.

Mario Piazza
University of Chieti-Pescara
Italy
E-mail: m.piazza@unich.it

Gabriele Pulcini
University of Cagliari
Italy
E-mail: gabriele.pulcini@unica.it

Intensionalisation of Logical Operators

Vít Punčochář[1]

Abstract: This paper introduces and explores semantics for propositional language in which every logical operator is intensional in a specific sense: its meaning is not based on truth conditions but on assertibility conditions. Assertibility is a relation between contexts and formulas, and contexts are modeled as sets of possible worlds. In the semantics, the consequence relation preserves assertibility instead of truth. A system of natural deduction for this semantics is formulated and completeness proved. At the end, the semantics is compared with two similar frameworks: Wansing's constructive connexive logic and inquisitive semantics.

Keywords: semantics, assertibility, intensionality, logical operators, natural deduction

1 Introduction

Entailment can be viewed as a relation which preserves some specific value. According to the standard picture the value in question is truth. This paper investigates an alternative: The consequence relation will be defined as assertibility preservation. This interpretation of the consequence relation is inspired mainly by Gauker (2005).

Assertibility of a given sentence is relative to a context in the same sense in which truth is relative to a possible world. So the first thing we will need is a formal representation of contexts. We will work with Stalnaker's concept of context: Contexts are represented simply as sets of possible worlds and possible worlds can be identified with classical valuations assigning either truth or falsity to every atomic formula (Stalnaker, 1999). This is the crucial difference between the approach taken in this paper and Gauker's approach taken in (Gauker, 2005). In his semantics, Gauker intentionally avoided the concept of possible world.

Since the semantic conditions determining the behaviour of negation, disjunction, conjunction, and implication will not be defined with respect to

---
[1] Work on this paper was supported by Grant No. 13-21076S of the Czech Science Foundation.

individual possible worlds but with respect to sets of possible worlds, the semantics will be called "semantics of intensionalized operators" (SIO).

The structure of this paper is as follows: In section 2, SIO will be introduced and its semantic conditions will be shortly motivated. The basic semantic concepts in SIO are the concepts of assertibility and deniability. Truth and falsity can be viewed as limiting cases of these concepts. That is the topic of section 3. SIO has several unusual properties that will be discussed in section 4. A system of natural deduction for SIO is introduced in section 5 and completeness is proved. The relation of SIO to two similar frameworks (Wansing's constructive connexive logic C and inquisitive semantics) is discussed in section 6.

2 Semantics of intensionalized operators

In this section, the semantics of intensionalized operators will be formulated and some motivation will be provided.

Contexts are nonempty sets of possible worlds understood as classical valuations, i.e., as functions from atomic formulas to truth values $\{0,1\}$. For the language of classical propositional logic (using the connectives $\neg$, $\vee$, $\wedge$, $\rightarrow$), we are going to define the relation of assertibility ($\Vdash^+$) between contexts and formulas. We will need also a parallel deniability relation ($\Vdash^-$) since assertibility of a negation will be defined as deniability of the negated formula. We will suppose that assertibility is on a par with deniability similarly as for example in the Nelson's logic $N4$ (see Odintsov, 2008) and Wansing's constructive connexive logic C which is a modification of $N4$ (see Wansing, 2005). In the following conditions, p ranges over atomic formulas and ϕ, ψ over all formulas of the propositional language.

A1 $C \Vdash^+ p$ iff for all $v \in C$, $v(p) = 1$.

D1 $C \Vdash^- p$ iff for all $v \in C$, $v(p) = 0$.

A2 $C \Vdash^+ \neg\phi$ iff $C \Vdash^- \phi$.

D2 $C \Vdash^- \neg\phi$ iff $C \Vdash^+ \phi$.

A3 $C \Vdash^+ \phi \vee \psi$ iff $C \Vdash^+ \phi$ or $C \Vdash^+ \psi$.

D3 $C \Vdash^- \phi \vee \psi$ iff $C \Vdash^- \phi$ and $C \Vdash^- \psi$.

A4 $C \Vdash^+ \phi \wedge \psi$ iff $C \Vdash^+ \phi$ and $C \Vdash^+ \psi$.

D4 $C \Vdash^- \phi \wedge \psi$ iff $C \Vdash^- \phi$ or $C \Vdash^- \psi$.

A5 $C \Vdash^+ \phi \to \psi$ iff $D \Vdash^+ \psi$ for all $D \subseteq_\emptyset C$ such that $D \Vdash^+ \phi$.

D5 $C \Vdash^- \phi \to \psi$ iff $D \Vdash^- \psi$ for all $D \subseteq_\emptyset C$, such that $D \Vdash^+ \phi$.

$D \subseteq_\emptyset C$ means that D is a nonempty subset of C. The symbol $\vDash$ will stand for the consequence relation of SIO which is defined as preservation of assertibility. By this definition SIO determines a particular logic (a set of formulas and inferential patterns which are regarded as valid) which will be called LIO (logic of intensionalized operators).

Now we will provide some motivation for the individual semantic conditions.

The conditions A1 and D1 for atomic formulas are motivated by the idea that p is assertible in a context iff there is enough evidence in the context that p is true. The context is just a set of possible worlds and it is supposed that the actual world is one of the worlds contained in the context but it is not determined which one it is. So there is enough evidence in the context that p is true iff p is true in all the worlds of the context. Similarly, there is enough evidence that p is false iff p is false in all the worlds.

The assertion "It is not the case that A" amounts to the denial of A. This view leads directly to the condition A2. The condition D2 is symmetric to A2.

The assertibility condition for disjunction requires special attention. The disjunction in SIO is a constructive one. For example, Christopher Gauker noticed that it is often appropriate to use disjunction in a constructive way when it connects two conditional sentences (see Gauker, 2005). The following inference taken from (Adams, 1975) may serve for an illustration.

> If switches A and B are thrown the motor will start. Therefore, either if switch A is thrown the motor will start or if switch B is thrown the motor will start.

This argument is obviously invalid. The problem is evident when we read "or" in the conclusion constructively: The conclusion holds with respect to a given context only if at least one of its disjuncts does. In correspondence with this informal example, $(p \to r) \vee (q \to r)$ does not follow from $(p \wedge q) \to r$ in LIO. It is easy to construct a formal counterexample: Take a context in which the intersection of the set of p-worlds (i.e. worlds which assign the value 1 to p) and the set of q-worlds is a subset of the set of r-worlds but neither the set of p-worlds nor the set of q-worlds is a subset of the set of r-worlds.

The condition A4 does not need any comment and the conditions D3 and D4 are based on the idea that deniability is on a par with assertibility.

The condition A5 says that implication functions as a kind of strict indicative conditional. It is indicative and not subjunctive in the sense that it respects the boundaries of the context: When evaluating a conditional formula in C, we can concentrate only on the subcontexts of C. It is strict in the sense that in order for $\phi \to \psi$ to be assertible, ψ has to be assertible in *all* (and not only some selected) ϕ-subcontexts (i.e. subcontexts in which ϕ is assertible). In particular, the formula $p \to q$ is assertible in a context iff the corresponding material implication is true in every possible world of the context.

However, as a strict implication, the arrow behaves quite untypically. For example, the so called "paradoxes of material implication" $p \to (q \to p)$ and $\neg p \to (p \to q)$ are assertible in every context and so valid according to the semantics. On the other hand, the implication avoids some drawbacks of the standard accounts of strict implication concerning especially the case of nested conditionals. For example, as the reader can easily verify, $p \to (q \to r)$ is equivalent to $(p \wedge q) \to r$ in LIO. This seems to be much more intuitive than how the formula $p \to (q \to r)$ is semantically evaluated when $\to$ is interpreted as, for example, the strict implication of the logic $S5$.

The motivation for D5 is based on the observation that, using Grice's words, "sometimes a denial of a conditional is naturally taken as a way of propounding a counterconditional, the consequent of which is the negation of the consequent of the original conditional. If A says 'If he proposes to her, she will refuse him' and B says 'That is not the case,' B would quite naturally be taken to mean 'If he proposes to her, she will not refuse him'" (Grice, 1989, p. 80). Technically speaking, this amounts to the equivalence between the formula $\neg(p \to q)$ and $p \to \neg q$ which is guaranteed by the condition D5. This equivalence plays a crucial role also in the context of the so called connexive logic (see Wansing, 2014). In fact, SIO is very close to Wansing's constructive connexive logic C introduced in (Wansing, 2005). The relation will be discussed later.

3 Assertibility vs. truth

In SIO, the basic semantic concepts are the concepts of assertibility and deniability in a context. The concepts of truth and falsity in a possible world can be defined as limiting cases of assertibility and deniability. Let us define that a formula ϕ is true in a possible world v iff it is assertible in the context $\{v\}$ and it is false in v iff it is deniable in $\{v\}$.

Note that then the logic of truth and falsity is not classical logic even though the singular valuations are classical. The condition D5 makes the difference. For a given world v, we obtain the following conditions of truth and falsity:

T1 p is true in v iff $v(p) = 1$.

F1 p is false in v iff $v(p) = 0$.

T2 $\neg \phi$ is true in v iff ϕ is false in v.

F2 $\neg \phi$ is false in v iff ϕ is true in v.

T3 $\phi \vee \psi$ is true in v iff ϕ is true in v or ψ is true in v.

F3 $\phi \vee \psi$ is false in v iff ϕ is false in v and ψ is false in v.

T4 $\phi \wedge \psi$ is true in v iff ϕ is true in v and ψ is true in v.

F4 $\phi \wedge \psi$ is false in v iff ϕ is false in v or ψ is false in v.

T5 $\phi \rightarrow \psi$ is true in v iff ϕ is not true in v or ψ is true in v.

F5 $\phi \rightarrow \psi$ is false in v iff ϕ is not true in v or ψ is false in v.

These semantic conditions remind the conditions of the so called material connexive logic described and axiomatized in (Wansing, 2014). However, there is one important difference. The conditions of material connexive logic are not relative to classical valuations and, as a result, the principle of excluded middle does not hold in that logic. It is immediately obvious that in SIO, the formula $p \vee \neg p$ is true in every possible world and it is easy to verify by induction that this holds even if p is replaced with any complex formula. Let us mention without proof that the logic of truth and falsity determined by SIO could be axiomatized by adding the scheme $\phi \vee \neg \phi$ to the axiomatic system of material connexive logic.

The main goal of this paper is to characterize the logic of assertibility and deniability. We will provide a syntactic characterization of this logic in section 5. But before that some unusual properties of this logic will be discussed.

4 Some unusual features of LIO

The logic of intensionalized operators has some unusual logical features. First of all, similarly to Wansing's constructive (and also material) connexive logic, it can be viewed as a paraconsistent logic which contains some contradictions (i.e. formulas of the form $\phi \wedge \neg\phi$) among its valid formulas (i.e. formulas which are assertible in every context). For instance, the formula $((p \wedge \neg p) \rightarrow \phi) \wedge \neg((p \wedge \neg p) \rightarrow \phi)$ is logically valid for any formula ϕ. This particular case can be seen as reflecting the fact, that the statements of the form "A is implied by a contradiction" have a problematic status.

Second, the set of logically valid formulas (i.e. formulas which are assertible in every context) is not closed under uniform substitution. For example, the formula $(p \rightarrow (q \vee r)) \rightarrow ((p \rightarrow q) \vee (p \rightarrow r))$ is logically valid but when we substitute $q \vee r$ for p we receive the formula $((q \vee r) \rightarrow (q \vee r)) \rightarrow (((q \vee r) \rightarrow q) \vee ((q \vee r) \rightarrow r))$ which is not logically valid.

Third, if "logically equivalent" means "assertible in the same contexts", then the law of universal interchangeability of logically equivalent formulas fails to hold. The law says that if two formulas ϕ and ψ are logically equivalent and an occurrence of ϕ in a formula χ is replaced with ψ, then the result of this replacement is logically equivalent to χ. Consider the following example. An atom p is logically equivalent to $(p \vee \neg q) \wedge (p \vee q)$ but $\neg p$ is not equivalent to $\neg((p \vee \neg q) \wedge (p \vee q))$ since the latter formula is equivalent to $(\neg p \wedge q) \vee (\neg p \wedge \neg q)$ which is not assertible in the context $\{v, w\}$ where $v(p) = 0, v(q) = 1, w(p) = 0, w(q) = 0$. Of course, in this context $\neg p$ is assertible.

Despite the fact that LIO does not satisfy the law of universal interchangeability, the semantics is compositional.[2] Besides logical equivalence we can define a notion of strong equivalence. We say that two formulas are strongly equivalent iff they are not only assertible but also deniable in the same contexts. In other words, two formulas are strongly equivalent if they are logically equivalent and their negations are logically equivalent as well. We will use the symbol $\rightleftharpoons$ for the relation of strong equivalence.

Fact 1 *If ϕ and ψ are strongly equivalent formulas and an occurrence of ϕ in a formula χ is replaced with ψ, then the result of the replacement is strongly equivalent to χ.*

---

[2]This phenomenon is analogous to the situation in the logics N4 and C.

Intensionalisation of Logical Operators

Proof. It is sufficient to observe that from the assumption $\phi \rightleftharpoons \psi$ it follows that $\neg\phi \rightleftharpoons \neg\psi$ and for an arbitrary formula χ

$$\phi \wedge \chi \rightleftharpoons \psi \wedge \chi \qquad\qquad \chi \wedge \phi \rightleftharpoons \chi \wedge \psi$$

$$\phi \vee \chi \rightleftharpoons \psi \vee \chi \qquad\qquad \chi \vee \phi \rightleftharpoons \chi \vee \psi$$

$$\phi \to \chi \rightleftharpoons \psi \to \chi \qquad\qquad \chi \to \phi \rightleftharpoons \chi \to \psi$$

$\square$

5 System of natural deduction for LIO

In this section, we will construct a Fitch style system of natural deduction for LIO and prove its weak completeness. The completeness proof exploits some ideas and techniques used also in (Punčochář, 2012).

The system consists of three groups of rules. The first group consists of rules for positive intuitionistic logic:

$(\wedge I)\quad \phi, \psi / \phi \wedge \psi \qquad (\wedge E)\quad$ (i) $\phi \wedge \psi / \phi$,
$\qquad\qquad\qquad\qquad\qquad\qquad\quad$ (ii) $\phi \wedge \psi / \psi$

$(\vee I)\quad$ (i) $\phi / \phi \vee \psi$,
$\qquad\quad$ (ii) $\psi / \phi \vee \psi \qquad (\vee E)\quad \phi \vee \psi, [\phi : \chi], [\psi : \chi] / \chi$

$(\to I)\quad [\phi : \psi]/\phi \to \psi \qquad (\to E)\quad \phi, \phi \to \psi / \psi$

The second group consists of rules for negated complex formulas:

$(\neg\wedge I)\quad \neg\phi \vee \neg\psi / \neg(\phi \wedge \psi) \qquad (\neg\wedge E)\quad \neg(\phi \wedge \psi)/\neg\phi \vee \neg\psi$
$(\neg\vee I)\quad \neg\phi \wedge \neg\psi / \neg(\phi \vee \psi) \qquad (\neg\vee E)\quad \neg(\phi \vee \psi)/\neg\phi \wedge \neg\psi$
$(\neg\to I)\quad \phi \to \neg\psi / \neg(\phi \to \psi) \qquad (\neg\to E)\quad \neg(\phi \to \psi)/\phi \to \neg\psi$
$(\neg\neg I)\quad \phi / \neg\neg\phi \qquad\qquad\qquad\qquad (\neg\neg E)\quad \neg\neg\phi/\phi$

The rules from the first and second group provide a complete system of rules for Wansing's connexive logic C. To receive a complete system for the logic LIO we need two more rules which can be applied only to a special type of formulas. We say that a formula α is simple iff it is built out of literals only by conjunction and implication. In the following two rules, α ranges over simple formulas and ϕ, ψ over all formulas.

$(ST)\qquad\quad / \alpha$ for every α provable in classical logic
$(\to \vee)\quad \alpha \to (\phi \vee \psi)/(\alpha \to \phi) \vee (\alpha \to \psi)$

⊢ will stand for the provability relation based on this system. The rule (ST) is quite untypical. It says that every simple theorem of classical logic can be inferred immediately in our system. As a result, the whole system depends on a system for classical logic.

Now, we are going to prove completeness. We will start with some observations concerning SIO.

Lemma 1 *If a formula is assertible in a context, then it is assertible in all subcontexts of the context.*

Proof. The claim can be easily proved by induction on the complexity of the formula. □

A logic has disjunction property if the following holds: Whenever $\phi \vee \psi$ is valid in the logic then ϕ is valid or ψ is valid.

Lemma 2 *LIO has disjunction property.*

Proof. Suppose that neither ϕ nor ψ is assertible in all contexts. Then there is a context C_1 in which ϕ is not assertible and there is a context C_2 in which ψ is not assertible. It follows from lemma 1 that $\phi \vee \psi$ is not assertible in $C_1 \cup C_2$. □

In lemma 3, the following notation will be used: Suppose that v is a possible world (i.e. a classical valuation) and α is a formula. $v(\alpha) = 1$ will mean that α is true in v according to the classical logic.

Lemma 3 *Suppose that α is a simple formula and C is a context. Then $C \Vdash^+ \alpha$ iff for all $v \in C$, $v(\alpha) = 1$.*

Proof. We can proceed by induction on the complexity of the simple formula. The case of literals is straightforward. We have to prove the inductive steps for conjunction and implication. Suppose that the claim holds for some simple formulas β and γ.

Conjunction: $C \Vdash^+ \beta \wedge \gamma$ iff $C \Vdash^+ \beta$ and $C \Vdash^+ \gamma$ iff for all $v \in C$, $v(\beta) = 1$ and for all $v \in C$, $v(\gamma) = 1$ iff for all $v \in C$, $v(\beta \wedge \gamma) = 1$.

Implication: $C \Vdash^+ \beta \to \gamma$ iff $D \Vdash^+ \gamma$ for all $D \subseteq_\emptyset C$, such that $D \Vdash^+ \beta$ iff for all $D \subseteq_\emptyset C$, such that $v(\beta) = 1$ for all $v \in D$, it holds that $v(\gamma) = 1$ for all $v \in D$ iff for all $v \in C$, if $v(\beta) = 1$, then $v(\gamma) = 1$ iff for all $v \in C$, $v(\beta \to \gamma) = 1$. □

Intensionalisation of Logical Operators

The following lemma is a direct consequence of lemma 3. The implication from left to right will be used in the proof of completeness and the implication from right to left in the proof of soundness.

Lemma 4 *A simple formula is assertible in every context iff it is a classical tautology.*

Now we can turn to the natural deduction system. For our completeness proof, the following lemma is crucial. $\phi \equiv \psi$ is an abbreviation for $\phi \vdash \psi$ and $\psi \vdash \phi$.

Lemma 5 *For every formula ϕ, there are simple formulas $\alpha_1, \ldots, \alpha_n$ ($n \geq 1$) such that $\phi \equiv \alpha_1 \vee \ldots \vee \alpha_n$.*

Proof. We will proceed by induction on the complexity of ϕ. Every literal is a simple formula. So we can suppose that the claim is true for some formulas ψ, χ and for their negations. So we assume that

$$\psi \equiv (\alpha_1 \vee \ldots \vee \alpha_k) \, (k \geq 1),$$

$$\chi \equiv (\beta_1 \vee \ldots \vee \beta_l) \, (l \geq 1),$$

$$\neg\psi \equiv (\gamma_1 \vee \ldots \vee \gamma_m) \, (m \geq 1),$$

$$\neg\chi \equiv (\delta_1 \vee \ldots \vee \delta_n) \, (n \geq 1),$$

where $\alpha_1, \ldots, \alpha_k, \beta_1, \ldots, \beta_l, \gamma_1, \ldots, \gamma_m, \delta_1, \ldots, \delta_n$ are simple formulas. As regards the inductive steps, we have to prove that the claim holds for $\psi \wedge \chi, \psi \vee \chi, \psi \to \chi$ and the negations of these formulas and also for the formula $\neg\neg\psi$.

The case of double negation is trivial. We just apply the rules $(\neg\neg I)$ and $(\neg\neg E)$. The cases of disjunction and conjunction are obtained with the help of the rules $(\wedge I), (\wedge E), (\vee I), (\vee E)$, which lead to the distributivity laws, and the rules $(\neg\wedge I), (\neg\wedge E), (\neg\vee I), (\neg\vee E)$.

The most complicated case is that of implication which will be explained in detail. The fact will be used that for arbitrary formulas ϕ, ψ and any simple formula α, the formula $\alpha \to (\phi \vee \psi)$ is provably equivalent to $(\alpha \to \phi) \vee (\alpha \to \psi)$. This holds due to the rules $(\to \vee), (\to I), (\to E)$, and $(\vee I)$. Moreover, for any formulas ϕ, ψ, χ, the formula $(\phi \vee \psi) \to \chi$ is provably equivalent to $(\phi \to \chi) \wedge (\psi \to \chi)$ (due to the rules $(\to I), (\to E), (\vee I), (\vee E)$ and $(\wedge E)$). As a consequence, equivalences between the following formulas can be established:

$$\psi \to \chi$$

$$(\alpha_1 \vee \ldots \vee \alpha_k) \to (\beta_1 \vee \ldots \vee \beta_l)$$

$$\bigwedge_{i=1}^{k}(\alpha_i \to (\beta_1 \vee \ldots \vee \beta_l))$$

$$\bigwedge_{i=1}^{k}((\alpha_i \to \beta_1) \vee \ldots \vee (\alpha_i \to \beta_l))$$

$$\bigvee_{i_1=1}^{l} \ldots \bigvee_{i_k=1}^{l}((\alpha_1 \to \beta_{i_1}) \wedge \ldots \wedge (\alpha_k \to \beta_{i_k}))$$

The case of negated conditional is similar. We just have to use the rules $(\neg \to I)$ and $(\neg \to E)$ so that we work with $\psi \to \neg \chi$ instead of $\psi \to \chi$ and β's are replaced with δ's. □

Theorem 1 $\phi_1, \ldots, \phi_n \vdash \psi$ iff $\phi_1, \ldots, \phi_n \vDash \psi$.

Proof. The implication from left to right requires the routine verification of soundness of all rules. The soundness of the rule (ST) is easily seen from lemma 4. We will prove the soundness of the rule $(\to \vee)$. Suppose that we are in a context C in which it is assertible $\alpha \to (\phi \vee \psi)$. It follows from lemma 3 that the union of the subcontexts in which α is assertible is a subcontext in which α is assertible. Let us call it D. $\phi \vee \psi$ is assertible in D which means that ϕ or ψ is assertible in D. It follows (due to lemma 1) that $\alpha \to \phi$ or $\alpha \to \psi$ is assertible in C.

We will prove the implication from right to left. Assume $\phi_1, \ldots, \phi_n \vDash \psi$. Then $\vDash (\phi_1 \wedge \ldots \wedge \phi_n) \to \psi$ since conditional proof is sound in LIO. Let χ denote the formula $(\phi_1 \wedge \ldots \wedge \phi_n) \to \psi$. Due to lemma 5, there are some simple formulas $\alpha_1, \ldots, \alpha_m$ ($m \geq 1$) such that $\chi \equiv \alpha_1 \vee \ldots \vee \alpha_m$. Since the system for LIO is sound, $\vDash \alpha_1 \vee \ldots \vee \alpha_m$. Since LIO has the disjunction property (lemma 2), there is some i ($1 \leq i \leq m$) such that $\vDash \alpha_i$. α_i is simple, therefore it is a tautology (lemma 4). Using the rule (ST), we can conclude that $\vdash \alpha_i$ and therefore also $\vdash \chi$. It follows that $\phi_1, \ldots, \phi_n \vdash \psi$. □

6 Comparisons

LIO has some common features with various logics known from the literature. In this section, two logical theories will be considered, which are, at least from the technical point of view, very close to LIO.

The first one is Wansing's constructive connexive logic C. It was already mentioned that the rules of the first and second group from the previous section constitute a system which is sound and complete with respect to the semantics of C. Therefore, the system for LIO extends a system for C by the rules (ST) and $(\rightarrow \vee)$.

The semantics of C is based on the class of C-models. A C-model is a structure $M = \langle W, \leq, V^+, V^- \rangle$ where W is a nonempty set, $\leq$ is a transitive and reflexive relation on W, and V^+, V^- are two valuations, i.e. functions from the set of atomic formulas to the set of upward closed subsets of W.[3] The semantic conditions correspond to the semantic conditions of SIO but they are relative to the individual members of W and instead of subcontexts, we have to look at the accessible worlds. In other words, $\leq$ in C functions as the superset relation $\supseteq$ in SIO.

SIO can be seen as the restriction of the semantics for C to a special class of C-models. In particular, the models are of the following structure: $\langle \wp(C) - \{\emptyset\}, \supseteq, V^+, V^- \rangle$, where C is a nonempty set of classical valuations, $\wp(C)$ is the powerset of C, $\supseteq$ is the superset relation, and V^+, V^- are defined in the following way:

$$V^+(p) = \{D \subseteq_\emptyset C; \text{for all } v \in D, v(p) = 1\}.$$

$$V^-(p) = \{D \subseteq_\emptyset C; \text{for all } v \in D, v(p) = 0\}.$$

These observations clarified how LIO is syntactically and semantically related to C. LIO can be easily related also with the so called inquisitive logic ($InqL$) and its semantics (see Ciardelli and Roelofsen 2011). Despite the fact that inquisitive semantics is usually interpreted in a way which differs significantly from the informal interpretation of SIO provided at the beginning of this paper, from the technical viewpoint these two semantics are very similar. The main difference concerns the semantic conditions for negation. In inquisitive semantics, the negation $\neg \phi$ is defined in the intuitionistic way as the implication of absurdity $\phi \rightarrow \bot$.

If we identify $InqL$ with the set of validities in inquisitive semantics and LIO as the set of validities in SIO, then it holds that the positive fragment of $InqL$ is identical with the positive fragment of LIO and the two logics can be interpreted one in the other with the help of the following two translations $*$ and $+$.

a) $p^* = p$ for every atomic formula p.

---

[3]These structures are also known as Nelson models. The semantic conditions of C and the Nelson's logic $N4$ are almost identical. The only difference makes the negative (deniability) condition for implication (see Wansing 2014).

b) $(\neg\phi)^* = \phi^* \to (p \wedge \neg p)$.

c) $(\phi \wedge \psi)^* = \phi^* \wedge \psi^*$.

d) $(\phi \vee \psi)^* = \phi^* \vee \psi^*$.

e) $(\phi \to \psi)^* = \phi^* \to \psi^*$.

Fact 2 $\phi \in \mathit{InqL}$ iff $\phi^* \in \mathit{LIO}$.

a) $p^+ = p$.

b) $(\neg p)^+ = \neg p$.

c) $(\neg\neg\phi)^+ = \phi^+$.

d) $(\phi \wedge \psi)^+ = \phi^+ \wedge \psi^+$.

e) $(\neg(\phi \wedge \psi))^+ = (\neg\phi)^+ \vee (\neg\psi)^+$.

f) $(\phi \vee \psi)^+ = \phi^+ \vee \psi^+$.

g) $(\neg(\phi \vee \psi))^+ = (\neg\phi)^+ \wedge (\neg\psi)^+$.

h) $(\phi \to \psi)^+ = \phi^+ \to \psi^+$.

i) $(\neg(\phi \to \psi))^+ = \phi^+ \to (\neg\psi)^+$.

Fact 3 $\phi \in \mathit{LIO}$ iff $\phi^+ \in \mathit{InqL}$.

These two facts show that the system of natural deduction from the previous section plus the translation * provide an indirect syntactic characterization of inquisitive logic and the axiomatic system for inquisitive logic (see Ciardelli and Roelofsen 2011) plus the translation + provide an indirect syntactic characterization of LIO.

7 Conclusion

The goal of this paper was to introduce and explore a particular semantic system for the language of classical propositional logic. The peculiarity of the system is that it is based on semantic conditions which are not relative to singular possible worlds but to sets of possible worlds (representing contexts) and so they are not truth (falsity) conditions but rather assertibility (deniability) conditions. Consequently, the semantics of all the operators (negation, implication, conjunction, and disjunction) is intensional in a special sense. The main result of this paper was a syntactic characterization of the logic determined by this semantics.

References

Adams, E. (1975). *The Logic of Conditionals. An Application of Probability to Deductive Logic*. Dordrecht: D. Reidel Publishing Company.
Ciardelli, I., & Roelofsen, F. (2011). Inquisitive Logic. *Journal of Philosophical Logic*, *40*, 55–94.
Gauker, C. (2005). *Conditionals in Context*. London: MIT Press.
Grice, H. (1989). *Studies In The Way Of Words*. London: Harvard University Press.
Odintsov, S. (2008). *Constructive Negations and Paraconsistency*. Dordrecht: Springer.
Punčochář, V. (2012). Some Modifications of Carnap's Modal Logic. *Studia Logica*, *100*, 517–543.
Stalnaker, R. (1999). *Context and Content*. Oxford: Oxford University Press.
Wansing, H. (2005). Connexive Modal Logic. In R. Schmidt, I. Pratt-Hartmann, M. Reynolds, & H. Wansing (Eds.), *Advances in Modal Logic* (Vol. 5, pp. 367–383). London: King's College Publications.
Wansing, H. (2014). Connexive Logic. In E. N. Zalta (Ed.), *The Stanford Encyclopedia of Philosophy* (Spring 2014 ed.).

Vít Punčochář
Institute of Philosophy, Czech Academy of Sciences
The Czech Republic
E-mail: vit.puncochar@centrum.cz

Three Floors for the Theory of Theory Change

Hans Rott[1]

Abstract: The theory of theory change due to Alchourrón, Gärdenfors and Makinson ("AGM") has been widely known as being characterised by two sets of postulates, one being very weak and the other being very strong. Commenting on the three classic constructions of partial meet contraction, safe contraction and entrenchment-based construction, I argue that three intermediate levels can be distinguished that play decisive roles within the AGM theory.

Keywords: theory change, partial meet contraction, safe contraction, entrenchment, interval orders, AGM

1 Introduction: The classic theory of AGM

The theory of theory change was sparked off by a seminal article by Carlos Alchourrón, Peter Gärdenfors, and David Makinson ("AGM") in the *Journal of Symbolic Logic* in 1985. While the original AGM theory in itself was certainly incomplete, it initiated a research program and has successfully been extended to the changes of theory bases (belief bases), to iterated and multiple belief change, non-prioritised belief revision, belief merging (belief fusion), and it has been embedded into modal logic frameworks in multi-agent contexts with higher-order beliefs.[2] In the present paper, I am not concerned with such annexes to the original AGM edifice. I will rather describe levels within the AGM theory that, I believe, have not been sufficiently recognised so far. AGM created, to use this metaphor, a large hall building with a floor and a ceiling. This edifice leaves a lot of room for designing the floor plan. I argue that the room may be structured by fitting in three intermediate floors that house some important incarnations of the theory of theory change.

---

[1] I am grateful to David Makinson for helpful comments on this paper.
[2] For excellent systematic and historical expositions, see (Hansson, 1999) and (Fermé & Hansson, 2011).

The AGM theory of the 1980s came in two packages that have most prominently been characterised by two sets of postulates. A set of six postulates constituted the basic theory, an extended set of eight postulates constituted the full theory. These postulates have become famous as the "AGM postulates" (sometimes also called the "Gärdenfors postulates"):

(K$\dotdiv$1)	if K is a theory, then $K\dotdiv\phi$ is a theory	(Closure)
(K$\dotdiv$2)	$K\dotdiv\phi \subseteq K$	(Inclusion)
(K$\dotdiv$3)	if $\phi \notin K$, then $K\dotdiv\phi = K$	(Vacuity)
(K$\dotdiv$4)	if $\phi \notin \mathrm{Cn}(\emptyset)$, then $\phi \notin K\dotdiv\phi$	(Success)
(K$\dotdiv$5)	$K \subseteq \mathrm{Cn}((K\dotdiv\phi) \cup \{\phi\})$	(Recovery)
(K$\dotdiv$6)	if $\mathrm{Cn}(\phi) = \mathrm{Cn}(\psi)$, then $K\dotdiv\phi = K\dotdiv\psi$	(Extensionality)
(K$\dotdiv$7)	$K\dotdiv\phi \cap K\dotdiv\psi \subseteq K\dotdiv(\phi\wedge\psi)$	(Conjunctive overlap)
(K$\dotdiv$8)	if $\phi \notin K\dotdiv(\phi\wedge\psi)$, then $K\dotdiv(\phi\wedge\psi) \subseteq K\dotdiv\phi$	(Conjunct. inclusion)

Justifications for these postulates are given in (Gärdenfors, 1988, pp. 61–65). It seems fair to say that it was only the addition of the supplementary postulates (K$\dotdiv$7) and (K$\dotdiv$8) that made the theory really interesting. It was recognised only somewhat later that the space between the basic and the full theory is huge and may be structured in various instructive ways. This understanding resulted from two developments that both began in about 1990. First, it became apparent that the theory of theory revision offers a framework for analysing theories of non-monotonic reasoning that had been discovered and rapidly developed in artificial intelligence research during the 1980s. Gärdenfors (1990) even suggested to think of belief revision—by then the most prominent application or interpretation of theory revision—and non-monotonic reasoning as "Two sides of the same coin." It became clear that most of the widely discussed principles of non-monotonic reasoning translate into principles for theory change that lie *between* the basic and the full AGM theory. Second, it was discovered that the theory of theory change (as well as the theory of non-monotonic reasoning) are amenable to a systematic interpretation in terms of rational choice (Lehmann, 2001; Lindström, 1991; Rott, 1993, 2001; Schlechta, 1997), and again, the major principles of rational choice all translate into principles for theory change that lie *between* the basic and the full AGM theory.

Even though the postulates are perhaps the most widely known part of the AGM theory, I do not think it is the postulates themselves that lend credibility and substance to the AGM theory. The success of the AGM program can be explained better by the fact that AGM at the same time developed three plausible constructive methods of revising theories and that these methods satisfy their postulates. In a way, the constructive methods served as a semantics for the postulates.

In this paper, I will work with an object language $\mathcal{L}$ the elements of which (the sentences) are denoted by lower-case Greek letters and the subsets of which are denoted by upper-case Latin letters. $\mathcal{L}$ contains the usual n-ary truth-functional operators $\bot$ and $\top$ (n=0), $\neg$ (n=1), $\vee$, $\wedge$, $\rightarrow$, and $\leftrightarrow$ (n=2). The logic governing $\mathcal{L}$ is a Tarskian consequence operation Cn that includes classical propositional logic, is reflexive, idempotent, monotonic, compact, and satisfies the deduction theorem. We write $M \vdash \phi$ for $\phi \in \mathrm{Cn}(M)$, $\phi \vdash \psi$ for $\psi \in \mathrm{Cn}(\{\phi\})$, $\phi \dashv\vdash \psi$ for $\mathrm{Cn}(\{\phi\}) = \mathrm{Cn}(\{\psi\})$, and $\vdash \phi$ for $\phi \in \mathrm{Cn}(\emptyset)$. By K and variants like K', $K \dotdiv \phi$ etc., we denote theories in $\mathcal{L}$, i.e., subsets of $\mathcal{L}$ that are closed under Cn. Theories are infinite sets, but some theories can be partitioned into finitely many equivalence classes with respect to Cn; such theories will be called *logically finite*.

We focus on the problem of *theory contraction*. Here the agent receives an input consisting of a sentence ϕ to be contracted from her theory K, and then performs an operation whose goal is to find a plausible outcome $K \dotdiv \phi$ that is a subset of the prior theory that does not imply ϕ. We now briefly describe the classical methods of theory contraction. We restrict the following definitions to the contraction of a non-tautological ϕ; for tautological sentences ϕ, $K \dotdiv \phi$ is put equal to K.

Partial meet contraction is based on a selection function (or choice function) σ which is applied to the set of maximal subsets of K that do not imply ϕ. These are called the ϕ-*remainders* of K; let $K \bot \phi$ be the set of ϕ-remainders of K. Then $\sigma(K \bot \phi)$ is a non-empty subset of $K \bot \phi$. Intuitively, that a remainder is in $\sigma(K \bot \phi)$ means that it belongs to those of $K \bot \phi$ that are most "plausible" or "secure" or "valuable".

Definition 1 *The partial meet contraction associated with σ is defined to be equal to the intersection of the selected remainders:*

$$K \dotdiv \phi = \bigcap \sigma(K \bot \phi)$$

We will later look at *relational partial meet contractions* $\dotdiv$ that can be constructed from a selection function σ that is based on a relation $\sqsubset$ on $K \bot \phi$, such that for all ϕ, $\sigma(K \bot \phi) = \{X \in K \bot \phi : X \sqsubset Y \text{ for no } Y \in K \bot \phi\}$.[3]

---

[3]Notice that $\sqsubset$ must be acyclic over $K \bot \phi$, as pointed out by (Alchourrón & Makinson, 1986, p. 189).

This equation defines the *rationalisation* of σ by $\sqsubset$; we also say that σ is based on *maximisation* with respect to $\sqsubset$. Grove (1988) showed that partial meet contractions can be transformed into contractions based on possible worlds models. It is important to realise that such models are *injective* (in the sense of (Freund, 1993)): Any two worlds can be distinguished by some sentence that is true at one but not at the other.

Safe contraction is based on a selection of sentences to be removed, rather than retained. The focus is on the set of minimal subsets of K that imply ϕ. These sets are called the ϕ-*kernels* of K; let $K \perp\!\!\!\perp \phi$ be the set of ϕ-kernels of K. In order to remove at least one element from each of the ϕ-kernels, a binary relation (a "hierarchy") $\prec$ on the elements of K is employed. Intuitively, $\psi \prec \chi$ means that ψ is less "secure or reliable or plausible" than χ (Alchourrón & Makinson, 1985, p. 411). From each ϕ-kernel, the minimally secure elements ψ, i.e., those ψ's for which there is no ρ in the kernel such that $\rho \prec \psi$, are selected for removal. A sentence ψ in K is $\prec$-*safe with respect to* ϕ if and only if it is not selected for removal in any of the ϕ-kernels. Let K/ϕ be the set of sentences in K that are safe with respect to ϕ. Then we have

Definition 2 *The safe contraction associated with $\prec$ is defined to be equal to the set of consequences of the safe elements:*

$$K \dotminus \phi = \mathrm{Cn}(K/\phi)$$

Like safe contraction, epistemic entrenchment-based contraction, or simply entrenchment contraction, is based on a binary relation $<$ on the elements of K. Here $\psi < \chi$ means that ψ is less "entrenched", or more easily given up, than χ.[4] Entrenchment contractions do not need a consideration of remainders or kernels, but rather operate by directly comparing certain elements of K.

Definition 3 *The entrenchment contraction associated with $<$ is defined to be*

$$K \dotminus \phi = \{\psi \in K : \phi < (\phi \vee \psi)\}$$

This construction has been standard within the full AGM theory since (Gärdenfors, 1988; Gärdenfors & Makinson, 1988), but it can be used in the much more general context of the basic AGM theory (Rott, 2003). Due to the occurrence of the disjunction, the definition of entrenchment contraction

---

[4]There is only a rather small difference between the intuitive readings of $\prec$ and $<$.

Three Floors for the Theory of Theory Change

is not very intuitive. It can be justified, however, by its perfect fit with the following reconstruction of comparative entrenchments from a reasoner's contraction behaviour:

Definition 4 *Assuming that a contraction function* $\dot{-}$ *over K is given, an entrenchment relation $<$ over K is defined by putting*

$$\phi < \psi \text{ iff } \psi \in K \dot{-} (\phi \wedge \psi) \text{ and } \nvdash \phi$$

It remains to recapitulate a number of properties of relations. A relation $<$ is acyclic iff there are no elements $x_1, \ldots, x_n$ such that $x_1 < \cdots < x_n < x_1$. A strict partial order is an irreflexive and transitive (and, thus, asymmetric) relation. An interval order is a strict partial order that satisfies the following *Interval condition*:

If $x < y$ and $u < v$, then either $x < v$ or $u < y$.

A strict partial order is modular (or "almost connected", or "virtually connected", or "negatively transitive") iff it satisfies the following condition:

If $x < y$, then either $x < z$ or $z < y$.

It is easy to check that for asymmetric relations, modularity implies the Interval condition, and that for irreflexive relations, the Interval condition in turn implies transitivity.

We now introduce some important properties that apply specifically to relations *over $\mathcal{L}$*:

(Extensionality)   If $\phi < \psi$, $\phi \dashv\vdash \phi'$ and $\psi \dashv\vdash \psi'$, then $\phi' < \psi'$.
(Continuing up)   If $\phi < \psi$ and $\psi \vdash \chi$, then $\phi < \chi$.
(Continuing down)   If $\phi < \psi$ and $\chi \vdash \phi$, then $\chi < \psi$.
(Choice easy)   If $\phi < \psi \wedge \chi$, then $\phi \wedge \psi < \chi$.
(Choice hard)   If $\phi \wedge \psi < \chi$ and $\phi \wedge \chi < \psi$, then $\phi < \psi \wedge \chi$.
(Maximality)   If $\nvdash \phi$, then $\phi < \psi$ for some ψ.
(K-Minimality)   If K is consistent, then ϕ is in K if and only if $\psi < \phi$ for some ψ.

Let us now introduce the notion of a hierarchy $\prec$ and of an entrenchment relation $<$ in a more precise way. The notion of a hierarchy for theory change is due to Alchourrón and Makinson (1985). We distinguish between three versions of the notion of entrenchment with increasing logical strength, introduced and motivated by Rott (2003), Rott (1992) and Gärdenfors and Makinson (1988), respectively.

Definition 5 *(a) A* hierarchy *is an acyclic relation over $\mathcal{L}$ that satisfies Extensionality.*
(b) A regular hierarchy *is a hierarchy that satisfies Continuing up and Continuing down.*
(c) A basic entrenchment relation *is an irreflexive relation over $\mathcal{L}$ that satisfies Extensionality, Choice easy, Choice hard and Maximality.*[5]
(d) A generalised entrenchment relation *is a basic entrenchment relation that satisfies Continuing up and Continuing down.*[6]
(e) A GM entrenchment relation *is a generalised entrenchment relation that is modular and satisfies K-Minimality.*

Continuing up and Continuing down (and thus Choice easy) are very plausible conditions for hierarchies and entrenchments. They encode that these notions are content-oriented and not completely unconstrained, like for instance judgements concerning the comparative reliability of the respective information sources.[7] The two Choice conditions are the hallmark of the notion of entrenchment. Notice that basic entrenchment relations are not in general acyclic, but generalised entrenchment relations are.[8] For a more thorough discussion of the connection between hierarchies and entrenchments, the reader is referred to (Rott & Hansson, 2014).

2 Three intermediate floors

Consider the following postulates.

(K$\dot{-}$8c) If $\psi \in K\dot{-}(\phi \wedge \psi)$, then $K\dot{-}(\phi \wedge \psi) \subseteq K\dot{-}\phi$.

---

[5]In contrast to the relations used in partial meet contraction and safe contraction (of logically finite theories), basic entrenchment relations need not be acyclic. But they are asymmetric. Suppose for reductio that $\phi < \psi$ and $\psi < \phi$. Then by Extensionality and Choice easy, $\phi \wedge \psi \wedge \phi < \psi$ and $\phi \wedge \psi \wedge \psi < \phi$. Thus by Choice hard $\phi \wedge \psi < \phi \wedge \psi$, contradicting Irreflexivity.

[6]Generalised entrenchment relations are transitive (and thus acyclic): Suppose that $\phi < \psi$ and $\psi < \chi$. Then by Continuing down $\phi \wedge \chi < \psi$ and $\phi \wedge \psi < \chi$, and by Choice hard $\phi < \psi \wedge \chi$, and thus by Continuing up $\phi < \chi$.

[7]For instance, the "prioritisations" in (Rott, 2009) are weak orderings over $\mathcal{L}$ that are not content-related.

[8]See (Rott, 1992, Lemma 4(v)) and (Rott, 2003, pp. 268–269).

(K$\dot{-}$8r) $K\dot{-}(\phi \wedge \psi) \subseteq \text{Cn}(K\dot{-}\phi \cup K\dot{-}\psi)$.
(K$\dot{-}$8p) If $\phi \in K\dot{-}(\phi \wedge \psi \wedge \chi)$, then $\phi \in K\dot{-}(\phi \wedge \psi)$ or $\phi \in K\dot{-}(\phi \wedge \chi)$.
(K$\dot{-}$8d) $K\dot{-}(\phi \wedge \psi) \subseteq K\dot{-}\phi \cup K\dot{-}\psi$.

Given the basic AGM postulates, these are all weakenings of (K$\dot{-}$8). (K$\dot{-}$8c) corresponds to Cumulative monotony (also known as "Cautious monotony") in non-monotonic reasoning and plays a central role in (Rott, 1992).[9] (K$\dot{-}$8r) was called (K$\dot{-}$8vwd) in (Rott, 2001); it is logically independent of (K$\dot{-}$8c), even for a logically finite K and in the presence of postulates (K$\dot{-}$1)–(K$\dot{-}$7).[10] Given the basic postulates, (K$\dot{-}$8p)[11] is strictly stronger than (K$\dot{-}$8r), and (K$\dot{-}$8d) in turn is strictly stronger than (K$\dot{-}$8p).[12] (K$\dot{-}$8d) corresponds to Disjunctive rationality in non-monotonic reasoning.[13] Given the basic postulates, (K$\dot{-}$8c) and (K$\dot{-}$8p) taken together imply (K$\dot{-}$8d); there is no analogous implication starting from (K$\dot{-}$8c) and (K$\dot{-}$8r).[14]

We keep the ground floor and the top floor established by AGM, and identify the following ones as intermediate floors:

1st floor: basic AGM plus (K$\dot{-}$7) and (K$\dot{-}$8r)
2nd floor: basic AGM plus (K$\dot{-}$7), (K$\dot{-}$8c) and (K$\dot{-}$8r)
3rd floor: basic AGM plus (K$\dot{-}$7), (K$\dot{-}$8c) and (K$\dot{-}$8d)

---

[9]The dual of (K$\dot{-}$8c) is (K$\dot{-}$7c): If $\psi \in K\dot{-}(\phi \wedge \psi)$, then $K\dot{-}\phi \subseteq K\dot{-}(\phi \wedge \psi)$. This is a weakening of (K$\dot{-}$7), and it corresponds to the Cut condition in non-monotonic reasoning.

[10]Cf. (Rott, 1993, p. 1438). The letter 'r' in (K$\dot{-}$8r) stands for 'relational', 'vwd' in (K$\dot{-}$8vwd) for 'very weak disjunctive'. The analogue of (K$\dot{-}$8r) for non-monotonic logic is important to (Freund, 1993) for characterising injective models in the logically finite case; it receives the name 'Weak disjunctive rationality' and takes centre stage in (Pino Pérez & Uzcátegui, 2000). The same name is used for a condition equivalent to (K$\dot{-}$8p) in (Rott, 2001).

[11]The letter 'p' is a somewhat accidental relic from (Rott, 1992) where it stood for 'preferential' or 'partial antitony'. The dual of (K$\dot{-}$8p), viz. (K$\dot{-}$7p): If $\phi \in K\dot{-}(\phi \wedge \psi)$, then $\phi \in K\dot{-}(\phi \wedge \psi \wedge \chi)$, is equivalent to (K$\dot{-}$7), given the basic postulates. On the choice-theoretic interpretation of (Rott, 2001), (K$\dot{-}$7p) and (K$\dot{-}$8p) correspond fairly directly to Sen's property α and (a finite version of) Sen's property γ, respectively. If a selection function takes all and only the finite subsets of some domain as arguments, then properties α and γ are necessary and jointly sufficient for the selection function to be relationalisable (or "rationalisable"). Cf. (Moulin, 1985) and (Sen, 1986, p. 1097).

[12]For the former claim, see Theorem 2, parts (ii) and (v), in (Rott & Hansson, 2014). For the latter claim, see the Appendix below.

[13]See (Rott, 2001, especially p. 104). The condition (K$\dot{-}$8d) is equivalent with a seemingly stronger one: $K\dot{-}(\phi \wedge \psi) \subseteq K\dot{-}\phi$ or $K\dot{-}(\phi \wedge \psi) \subseteq K\dot{-}\psi$. This condition is called the "Covering" condition in (Alchourrón, Gärdenfors, & Makinson, 1985) and "Strong disjunctive rationality" in (Rott, 2001).

[14]See Theorem 2, parts (iv) and (v), of (Rott & Hansson, 2014).

From the above comments it should be clear that all these floors are above AGM's ground floor and below AGM's top floor, and that floors with higher numbers are above the ones with lower numbers. "Being above" is metaphorical here for "being logically stronger".

In the remainder of this paper I will explain why I suggest that these levels are the right ones to be added to the floor plan of AGM's edifice—at least as long as one focusses on logically finite theories.[15] On each of the three floors, the results pertaining to partial meet and safe contractions are restricted to the case of logically finite theories K. No such restriction applies to entrenchment contractions.

2.1 The ground floor

It has been known since (Alchourrón et al., 1985) that the partial meet contractions are exactly the ones that satisfy the basic postulates (K$\dot{-}$1)–(K$\dot{-}$6), and since (Rott, 2003) that the same is true for the basic entrenchment contractions. Alchourrón and Makinson (1985) demonstrated that safe contractions satisfy the basic postulates, but it is shown in (Rott & Hansson, 2014) that there are contraction operations that satisfy the basic postulates and cannot be represented as safe contractions. Intuitively, the former allow cycles in the contraction behaviour that are forbidden by the latter. Safe contractions live somewhere between the ground floor and the 1st floor.[16]

2.2 The 1st floor

It has been known since (Rott, 1993, Corollary 2) that for logically finite theories K, the relational partial meet contractions are exactly the ones that satisfy the postulates (K$\dot{-}$1)–(K$\dot{-}$7) and (K$\dot{-}$8r).

The central result of (Alchourrón & Makinson, 1986) implies that the same postulates characterise safe contractions of logically finite theories based on regular hierarchies. More precisely, Continuing down is sufficient for the first floor, but for any first-floor contraction function, the representing safe contraction can be chosen so that it is based on a hierarchy that satisfies in addition Continuing up, as well as

---

[15] (K$\dot{-}$p) will play a supporting role. It is interesting in itself only if (K$\dot{-}$8c) is not satisfied (or, what comes to much the same thing, only if the underlying preference relation is not transitive).

[16] Is is not known where exactly they live. (Rott & Hansson, 2014, p. 41) contains a conjecture that the axiom characterising safe contractions is (K$\dot{-}$Acyc): If $\phi_{i+1} \in K\dot{-}(\phi_i \wedge \phi_{i+1})$ and $\nvdash \phi_i \wedge \phi_{i+1}$ for all $i=1,\ldots,n-1$, then not both $\phi_1 \in K\dot{-}(\phi_1 \wedge \phi_n)$ and $\nvdash \phi_1 \wedge \phi_n$.

(EII) If $\phi \wedge \psi < \chi$, then either $\phi < \chi$ or $\psi < \chi$.[17]

For entrenchment contractions, (K$\dot{-}$7) corresponds exactly to Continuing down, while (K$\dot{-}$8r) corresponds exactly to the rather unintuitive

(EII$^-$) If $\phi \wedge \psi < \chi$ and $\not\vdash \chi$, then there are ξ and ρ such that $\chi \dashv\vdash \xi \wedge \rho$ and $\phi \wedge \rho < \xi$ and $\psi \wedge \xi < \rho$.[18]

As the name indicates, EII$^-$ is a weakening of EII.

2.3 The 2nd floor

It has been known since (Rott, 1993, Corollary 2) that for logically finite theories K, the transitively relational partial meet contractions are exactly the ones that satisfy the postulates (K$\dot{-}$1)–(K$\dot{-}$7), (K$\dot{-}$8c) and (K$\dot{-}$8r). A very similar result was proven for non-monotonic logics on logically finite languages by Freund (1993, Theorem 4.13). Just the route of approaching this result was different. For AGM reinterpreted via the Grove connection, all models are injective; then comes the relationality of selection functions (1st floor), and last comes their transitive relationality (2nd floor). In Freund's models, everything is transitively relational (which is why Preferential logic includes Cumulative monotony, the analogue of (K$\dot{-}$8c)); and injectiveness comes only at the end.

An inspection of the proof of the central result of (Alchourrón & Makinson, 1986) shows that (K$\dot{-}$1)–(K$\dot{-}$7), (K$\dot{-}$8c) and (K$\dot{-}$8r) also characterise safe contractions based on transitive and regular hierarchies. It is also pointed out in (Rott & Hansson, 2014, section 3.4) that for any second-floor contraction function, it is possible to choose a representing safe contraction that is based on a hierarchy satisfying in addition either EII or Choice hard—but in general not both of them.

For entrenchment contractions, the new condition (K$\dot{-}$8c) corresponds exactly to Continuing up. So the second floor is inhabited by generalised entrenchment relations that satisfy EII$^-$.

---

[17](Alchourrón & Makinson, 1985, Observations 4.3 and 5.3) and (Rott & Hansson, 2014, Theorems 8 and 13). EII follows from the conjunction of Extensionality, Choice hard and Modularity. For suppose that EII is violated, i.e., that there are ϕ, ψ and χ such that $\phi \wedge \psi < \chi$ and $\phi \not< \chi$ and $\psi \not< \chi$. Then by Modularity, $\phi \wedge \psi < \phi$ and $\phi \wedge \psi < \psi$. Thus by Choice hard and Extensionality, $\psi < \phi$ and $\phi < \psi$, contradicting Asymmetry.

[18]See (Rott, 2001, Observation 68) and (Rott & Hansson, 2014, Theorem 2 and Lemma 10). "Correspondence" here means "correspondence via Definitions 3 and 4".

2.4 The 3rd floor

It has been known at least since (Rott, 2001, Observations 25 and 26) that for all theories K, a partial meet contraction function satisfies (K$\dot{-}$8d) if and only if the selection function σ on which it is based satisfies the following condition:[19]

(II$^+$)   Either $\sigma(S) \subseteq \sigma(S \cup S')$ or $\sigma(S') \subseteq \sigma(S \cup S')$.

Now we can show that any finitary choice function (i.e., any choice function over the set of subsets of a finite domain—this is the type of domain relevant for contractions of logically finite theories), that satisfies (II$^+$) can be rationalised by an irreflexive relation that satisfies the Interval condition, and that conversely each choice function rationalisable in this way satisfies (II$^+$).[20] So the partial meet contractions living on the 3rd floor are the ones that are based on interval orders. Given the prominence of interval orders within the theory of preferences and utilities,[21] this justifies the introduction of a new level to our floor plan.

For entrenchment contractions, (K$\dot{-}$8d) corresponds exactly to

(EII$^+$)   If $\phi \wedge \psi < \chi \wedge \xi$ and $\not\vdash \chi$ and $\not\vdash \xi$, then either $\phi < \chi$ or $\psi < \xi$.

As the name indicates, EII$^+$ is a strengthening of EII. EII$^+$ is almost as unintuitive as EII$^-$. Fortunately, the situation is improved by the above-mentioned fact that given (K$\dot{-}$1)–(K$\dot{-}$7) and (K$\dot{-}$8c), (K$\dot{-}$8p) is equivalent

---

[19] This is actually a rewritten form of condition (II$^+$) as given in (Rott, 2001, p. 163). A very thorough discussion of Disjunctive rationality in non-monotonic reasoning can be found in (Freund, 1993). For the sake of completeness, I give two semantic conditions that correspond to (K$\dot{-}$p) and are not, I think, in the literature so far. The first corresponds directly to (K$\dot{-}$p): If $\sigma(S \cup S' \cup S'') \cap S = \emptyset$, then $\sigma(S \cup S') \cap S = \emptyset$ or $\sigma(S \cup S'') \cap S = \emptyset$. The second corresponds to the condition (K$\dot{-}$wd) which was proven equivalent to (K$\dot{-}$p) in (Rott & Hansson, 2014, Theorem 2(ii)): Either $\sigma(S) \cap S' \subseteq \sigma(S \cup S')$ or $\sigma(S') \cap S \subseteq \sigma(S \cup S')$. The corresponding constraint on relations rationalising such choice functions is this: If $x < y$ and $u < v$, then either $x < v$ or $u < y$ or $x < u$ or $u < x$. See Appendix.

[20] See Appendix. I have not seen this very simple characterisation of "interval choice", i.e., maximising choice based on interval orders, in the literature. The most well-known choice condition for interval choice was found by Fishburn (1975) and can be written in this way: Either $\sigma(S) \cap S' \subseteq \sigma(S')$ or $\sigma(S') \cap S \subseteq \sigma(S)$. It is sometimes called, a little confusingly, "functional asymmetry". If a finitary choice function σ satisfies Sen's properties α and γ and Aizerman's property (cf. Moulin, 1985), then Fishburn's condition is equivalent with (II$^+$).

[21] Interval orders make room for the idea of nontransitive indifference relations due to discrimination thresholds. This idea originated in psychology in 1860 (Gustav Th. Fechner) and in economics in the late 1930s (Nicholas Georgescu-Roegen, Wallace E. Armstrong). For further results, cf. (Fishburn, 1970, 1985; Luce, 1956; Suppes, Krantz, Luce, & Tversky, 1989, Chapter 16; Wiener, 1914). Rabinowicz (2008) criticizes the use of interval orders in value theory. Booth and Meyer (2011) utilise them to construct revisions of doxastic states.

to (K∸8d). So we can substitute (K∸8p) for (K∸8d) in our characterisation of the 3rd floor, and (K∸8p) corresponds exactly to the much nicer condition EII for entrenchments.[22]

It is shown in (Rott & Hansson, 2014) that on the third floor, the relations retrieved with the help of Definition 4, i.e., regular hierarchies that satisfy Choice hard and EII, can be used for safe contractions with the same result as for entrenchment contractions. Since the latter method guarantees the satisfaction of (K∸8d), we infer that safe contractions do so, too.

Makinson (1994, p. 97) seemed to favour living on the 3rd floor when he concluded, towards the end of his survey article on non-monotonic reasoning, that Disjunctive rationality is desirable but Rational monotony is too strong to insist upon.[23] Still we shall briefly visit the top floor.

2.5 The top floor

The top floor is made up of the full AGM theory which we take here as an ideal limit of rational involvement in theory change. Terminologically, this is in line with the linguistic usage in non-monotonic logics since (Kraus, Lehmann, & Magidor, 1990) where the analogue of (K∸8) was called "Rational monotony". In all three classic AGM models of theory change, the top floor is reached by adding Modularity to the conditions one had before. This was essentially part of the fully fledged AGM theory of the 1980s, but their partial meet and entrenchment contractions, which were originally defined in terms of non-strict relations $\leq$ (or likewise $\sqsubseteq$), need defining the corresponding strict relation as the converse of complement, i.e., by putting $x < y$ iff not $y \leq x$. A statement like $x \leq y$ in (Alchourrón et al., 1985) or in (Gärdenfors & Makinson, 1988) thus has to be taken as meaning "x is less than or equal to or incomparable with y" rather than "x is less than or equal to y" (which explains the marginal role of connectivity and the starring role of transitivity in these papers).

---

[22] See (Rott, 2001, Observations 35, 36 and 51) and (Rott & Hansson, 2014, Theorem 2).

[23] Makinson mentions the Interval condition as sufficient for Disjunctive rationality and conjectures that it characterises Disjunctive rationality. The Filtering condition of (Freund, 1993, p. 244), which is suitable for the characterisation of Disjunctive rationality in his framework, is weaker than the Interval condition.

3 Conclusion

The edifice of theory change built by AGM, which looks like a big hall building, needs signposts to intermediate floors in order to become more habitable. I have tried to show that three floors have been there all along, in the structure of AGM's constructive models. But they have hardly been recognised.

Motivation and valuable guidance for the finer structuring of this theory have come from the theory of rational choice and the theory of non-monotonic reasoning. Especially the latter can readily be compared with the theory of theory change. Since the beautiful work of Makinson (1989, 1994), Kraus et al. (1990), and Lehmann and Magidor (1992), we have been accustomed to the thought that the most important levels in non-monotonic logic are constituted by Cumulative logic (consisting of Reflexivity, Left logical equivalence, Right weakening, Cut and Cumulative monotony), Preferential logic (adding Or) and Rational logic (further adding Rational monotony).

Gärdenfors and Makinson's idea that non-monotonic reasoning might be reconstructed as a special application of the theory of theory change has always had a strong appeal. However, the architecture of their theory has now turned out to have a floor map that differs from that of non-monotonic reasoning. We have found no independent use for the analogue of the Cut condition; it is simply a consequence of $(K \dotminus 7)$, the analogue of Or which we met already on the first floor.[24] The first floor also houses $(K \dotminus 8r)$, the analogue of a weakened form of Disjunctive rationality which has no special status in non-monotonic reasoning. Cumulative logic is obtained only on the second floor which at the same time completes Preferential logic, accompanied by $(K \dotminus 8r)$ again. The idea of Disjunctive rationality resides in the third floor and appears to play a more important role within the AGM edifice than within non-monotonic logic. Only the top floor nicely mirrors the system of Rational reasoning that we know from non-monotonic logic. But it is not clear whether we should, as rational reasoners, aspire to ascend as high as this in everyday theory change.

---

[24] Cf. footnotes 9 and 11 above.

Three Floors for the Theory of Theory Change

Appendix: some figures and proofs

Claim 1 *(K÷8d) is strictly stronger than (K÷8p).*

Proof. We use the Grove connection in order to show this. Consider the language $\mathcal{L}$ with just two atoms p and q, and the inconsistent theory $K = \mathrm{Cn}(p \wedge \neg p)$. Let the relation $<$ between the four possible worlds for $\mathcal{L}$ be fully specified in Figure 1. It is essential that this relation is not

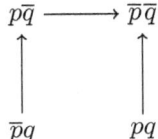

Figure 1: *A non-transitive ordering of models for $\mathcal{L}$. Arrows point to the less plausible models.*

transitive (its transitive closure is an interval order). The contraction function thus generated is represented in Figure 2. It is a tedious but simple

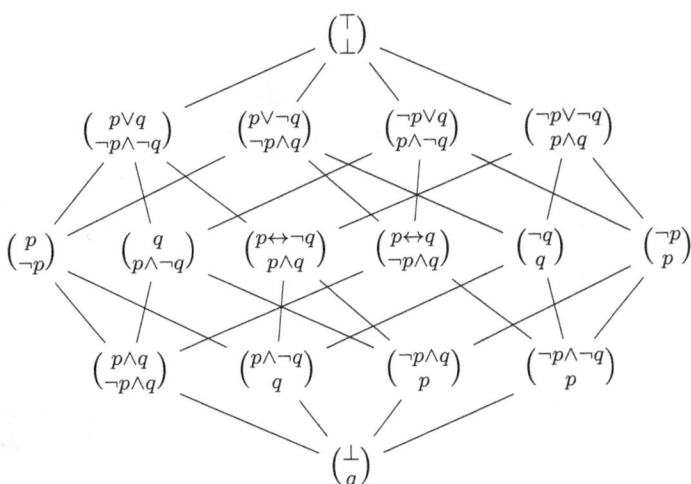

Figure 2: *The Hasse diagram for a contraction function over $K = K_\perp$ that satisfies (K÷1)–(K÷7), (K÷8r) and (K÷8p), but not (K÷8c) and not (K÷d). Read "$\binom{\phi}{\psi}$" as "$K \dot{-} \phi = \mathrm{Cn}(\psi)$".*

task to check, using maximisation $K \dot{-} \phi = \{\psi : \psi \in K$ and ψ is true at all w that are maximal among the worlds not satisfying $\phi\}$, that (K$\dot{-}$p) is satisfied; in particular, the fact that $q \in K\dot{-}(q \wedge p \wedge \neg p)$ is matched by $q \in K\dot{-}(q \wedge p) = \text{Cn}(\neg p \wedge q)$, as desired. However, (K$\dot{-}$d) is violated, since we obtain

$q \in K\dot{-}(p \wedge \neg p) = \text{Cn}(q)$, but neither $q \in K\dot{-}p = \text{Cn}(\neg p)$
nor $q \in K\dot{-}\neg p = \text{Cn}(p)$.

(K$\dot{-}$8c) is violated, too. We have

$q \in K\dot{-}(p \wedge q)$, but not $K\dot{-}(p \wedge q) \subseteq K\dot{-}p$,

because q is not in $K\dot{-}p$. Correspondingly, the entrenchment relation $<$ that is retrieved by Definition 4 violates Transitivity and Continuing up: We have

$p < q$ (and also $q < p \vee q$), but not $p < p \vee q$.

The full entrenchment structure for this example is given in Figure 3. Notice that there is no arrow from $p \vee q$ to p, and none from $p \vee q$ to $p \vee \neg q$. □

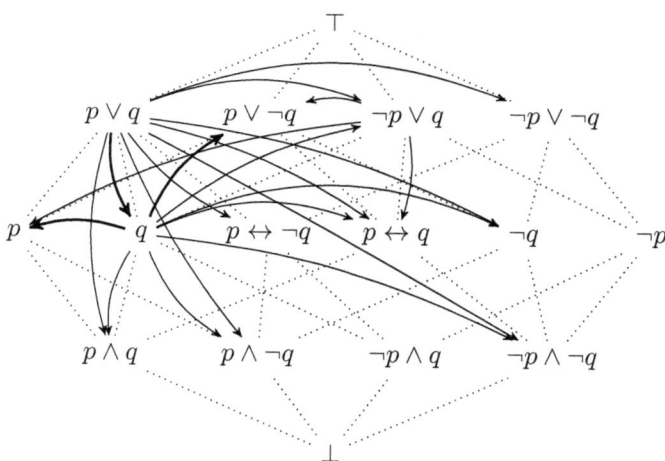

Figure 3: *Entrenchment diagram for the example of Figures 1 and 2. Arrows point to the less entrenched sentences. Arrows from and to $\top$ and $\bot$ are not drawn.*

Claim 2 *The following two semantic choice-theoretic conditions corresponding to (K$\dot{-}$p) and (K$\dot{-}$wd) respectively are equivalent. (See text to footnote 19.)*

Three Floors for the Theory of Theory Change 201

(II^p)   If $\sigma(S \cup S' \cup S'') \cap S = \emptyset$, then $\sigma(S \cup S') \cap S = \emptyset$
or $\sigma(S \cup S'') \cap S = \emptyset$.
(II^{wd})   Either $\sigma(S) \cap S' \subseteq \sigma(S \cup S')$ or $\sigma(S') \cap S \subseteq \sigma(S \cup S')$.

Proof. (II^p) implies (II^{wd}): Suppose that (II^p) is true, but (II^{wd}) is false. The latter says that
$$\sigma(S) \cap S' \not\subseteq \sigma(S \cup S') \text{ and } \sigma(S') \cap S \not\subseteq \sigma(S \cup S').$$
Using the abbreviation $X = (S \cap S') \setminus \sigma(S \cup S')$, a simple set-theoretic transformation tells us that this is equivalent to
$$\sigma(S \cup X) \cap X \neq \emptyset \text{ and } \sigma(S' \cup X) \cap X \neq \emptyset.$$
Then (II^p) yields $\sigma(S \cup S' \cup X) \cap X \neq \emptyset$. Putting back the definition of X again, we get $\sigma(S \cup S') \cap (S \cap S') \setminus \sigma(S \cup S') \neq \emptyset$. But this is obviously false, so our initial supposition cannot be true.

(II^{wd}) implies (II^p): Assume that (II^{wd}) is true. Then we get that either $\sigma(S \cup S') \cap (S \cup S'') \subseteq \sigma(S \cup S' \cup S'')$ or $\sigma(S \cup S'') \cap (S \cup S') \subseteq \sigma(S \cup S' \cup S'')$. Intersecting everything with S gives that either $\sigma(S \cup S') \cap S \subseteq \sigma(S \cup S' \cup S'') \cap S$ or $\sigma(S \cup S'') \cap S \subseteq \sigma(S \cup S' \cup S'') \cap S$. But this proves ($II^p$). □

Claim 3 *The constraint (II^{wd}) for the finitary and rationalisable semantic choice function σ that corresponds to ($K \dot{-} wd$) is equivalent with the following weakening of the Interval condition for the rationalising relation $<$ (cf. footnote 19):*

(II^{wd})   Either $\sigma(S) \cap S' \subseteq \sigma(S \cup S')$ or $\sigma(S') \cap S \subseteq \sigma(S \cup S')$.
($<^{wd}$)   If $x < y$ and $u < v$, then either $x < v$ or $u < y$ or $x < u$ or $u < x$.

Proof. Let σ be a finitary choice function that is rationalised by $<$.

(II^{wd}) implies ($<^{wd}$): Suppose that $x < y$ and $u < v$, and consider the sets $S = \{x, y, u\}$ and $S' = \{x, u, v\}$. Suppose for reductio that neither $x < v$ nor $u < y$ nor $x < u$ nor $u < x$. From $x \not< u$ and $x \not< v$, we get by maximisation that $x \in \sigma(S')$, and thus $x \in \sigma(S') \cap S$. Similarly, we get from $u \not< x$ and $u \not< y$ that $u \in \sigma(S) \cap S'$. Thus by (II^{wd}), either $x \in \sigma(S \cup S')$ or $u \in \sigma(S \cup S')$. But since by supposition $x < y$ and $u < v$, this contradicts maximisation.

($<^{wd}$) implies (II^{wd}): Suppose that neither $\sigma(S) \cap S' \subseteq \sigma(S \cup S')$ nor $\sigma(S') \cap S \subseteq \sigma(S \cup S')$. That is, there is an $x \in (\sigma(S) \cap S') \setminus \sigma(S \cup S')$, and a $y \in (\sigma(S') \cap S) \setminus \sigma(S \cup S')$. Since $x \notin \sigma(S \cup S')$, there must be a u in $S \cup S'$ such that $x < u$, and u is not in S, since $x \in \sigma(S)$. So $u \in S'$. Similarly, since $y \notin \sigma(S \cup S')$, there must be a v in $S \cup S'$ such that $y < v$, and v is not in S', since $y \in \sigma(S')$. So $v \in S$. From $x < u$ and $y < v$, we conclude with ($<^{wd}$) that either $x < v$ or $y < u$ or $x < y$ or $y < x$. But each of $x < y$ and $x < v$ contradict maximisation, since $x \in \sigma(S)$ and both y and v are in S. Similarly, each of $y < x$ and $y < u$ contradict maximisation, since $y \in \sigma(S')$ and both x and u are in S'. So all possibilities lead to a contradiction, which means that the supposition is incompatible with ($<^{wd}$). □

Claim 4 *The constraint (II$^+$) for the finitary and rationalisable semantic choice function σ that corresponds to (K$\dot{-}$d) is equivalent with the Interval condition for the rationalising relation $<$. (See text to footnote 20.)*

(II$^+$) Either $\sigma(S) \subseteq \sigma(S \cup S')$ or $\sigma(S') \subseteq \sigma(S \cup S')$.
(Interval) If $x < y$ and $u < v$, then either $x < v$ or $u < y$.

Proof. (II$^+$) implies (Interval): Suppose that $x < y$ and $u < v$; we want to show that either $x < v$ or $u < y$. By maximisation, the supposition implies that $\sigma(\{x, y\}) = \{y\}$, $\sigma(\{u, v\}) = \{v\}$ and $x, u \notin \sigma(\{x, y, u, v\})$. By the requirement that σ deliver non-empty subsets (or by (II$^+$)), it follows that either $y \in \sigma(\{x, y, u, v\})$ or $v \in \sigma(\{x, y, u, v\})$. Assume without loss of generality that $y \in \sigma(\{x, y, u, v\})$; the other case is similar. Then by maximisation, $y \in \sigma(\{y, u\})$. If $u \notin \sigma(\{y, u\})$, then again by maximisation $u < y$, as desired. If, on the other hand, $u \in \sigma(\{y, u\})$, then not $\sigma(\{y, u\}) \subseteq \sigma(\{x, y, u, v\})$. Hence, by (II$^+$), $\sigma(\{x, v\}) \subseteq \sigma(\{x, y, u, v\})$. Since $x \notin \sigma(\{x, y, u, v\})$, it follows that $x \notin \sigma(\{x, v\})$. By maximisation, it follows that $x < v$, as desired.

(Interval) implies (II$^+$): Let $<$ be an interval ordering and suppose for reductio that neither $\sigma(S) \subseteq \sigma(S \cup S')$ nor $\sigma(S') \subseteq \sigma(S \cup S')$. Then there are $x \in \sigma(S)$ and $y \in \sigma(S')$ such that $x, y \notin \sigma(S \cup S')$. By maximisation, this means that there are u and v in $S \cup S'$ such that $x < u$ and $y < v$. Thus, by (Interval), either $x < v$ or $y < u$. Assume without loss of generality that $x < v$; the other case is similar. By maximisation, $x \in \sigma(S)$ implies $v \notin S$. Thus $v \in S'$. But this contradicts $y < v$, $y \in \sigma(S')$ and maximisation. □

References

Alchourrón, C. E., Gärdenfors, P., & Makinson, D. (1985). On the Logic of Theory Change: Partial Meet Contraction and Revision Functions. *Journal of Symbolic Logic, 50*, 510–530.

Alchourrón, C. E., & Makinson, D. (1985). On the Logic of Theory Change: Safe Contraction. *Studia Logica, 44*, 405–422.

Alchourrón, C. E., & Makinson, D. (1986). Maps between Some Different Kinds of Contraction Function: The Finite Case. *Studia Logica, 45*, 187–198.

Booth, R., & Meyer, T. (2011). How to Revise a Total Preorder. *Journal of Philosophical Logic, 40*, 193–238.

Fermé, E., & Hansson, S. O. (2011). AGM 25 Years: Twenty-Five Years of Research in Belief Change. *Journal of Philosophical Logic, 40*, 295–331.

Fishburn, P. C. (1970). Intransitive Indifference in Preference Theory: A Survey. *Operations Research, 18*, 207–228.

Fishburn, P. C. (1975). Semiorders and Choice Functions. *Econometrica, 43*, 975–977.

Fishburn, P. C. (1985). *Interval Orders and Interval Graphs: A Study of Partially Ordered Sets*. New York: Wiley.

Freund, M. (1993). Injective Models and Disjunctive Relations. *Journal of Logic and Computation, 3*, 231–247.

Gärdenfors, P. (1988). *Knowledge in Flux: Modeling the Dynamics of Epistemic States*. Cambridge, MA: MIT Press.

Gärdenfors, P. (1990). Belief Revision and Nonmonotonic Logic: Two Sides of the Same Coin? In L. C. Aiello (Ed.), *9th European Conference on Artificial Intelligence (ECAI'90)* (pp. 768–773). London: Pitman.

Gärdenfors, P., & Makinson, D. (1988). Revisions of Knowledge Systems Using Epistemic Entrenchment. In M. Vardi (Ed.), *Proceedings of the Second Conference on Theoretical Aspects of Reasoning About Knowledge (TARK'88)* (pp. 83–95). Los Altos: Morgan Kaufmann.

Grove, A. (1988). Two Modellings for Theory Change. *Journal of Philosophical Logic, 17*, 157–170.

Hansson, S. O. (1999). *A Textbook on Belief Dynamics*. Dordrecht: Kluwer.

Kraus, S., Lehmann, D., & Magidor, M. (1990). Nonmonotonic Reasoning, Preferential Models and Cumulative Logics. *Artificial Intelligence*, *44*, 167–207.

Lehmann, D. (2001). Nonmonotonic Logics and Semantics. *Journal of Logic and Computation*, *11*, 229–256.

Lehmann, D., & Magidor, M. (1992). What Does a Conditional Knowledge Base Entail? *Artificial Intelligence*, *55*, 1–60.

Lindström, S. (1991). *A Semantic Approach to Nonmonotonic Reasoning: Inference Operations and Choice* (Tech. Rep. No. 1991:6). Uppsala: University of Uppsala. (Officially published as Tech. Rep. No. 1994:10)

Luce, R. D. (1956). Semiorders and a Theory of Utility Discrimination. *Econometrica*, *24*, 178–191.

Makinson, D. (1989). General Theory of Cumulative Inference. In M. Reinfrank, J. de Kleer, M. L. Ginsberg, & E. Sandewall (Eds.), *Nonmonotonic Reasoning: Proceedings of the 2nd International Workshop 1988* (pp. 1–18). Berlin: Springer.

Makinson, D. (1994). General Patterns in Nonmonotonic Reasoning. In D. M. Gabbay, C. J. Hogger, & J. A. Robinson (Eds.), *Handbook of Logic in Artificial Intelligence and Logic Programming* (Vol. 3, pp. 35–110). Oxford: Oxford University Press.

Moulin, H. (1985). Choice Functions over a Finite Set: A Summary. *Social Choice and Welfare*, *2*, 147–160.

Pino Pérez, R., & Uzcátegui, C. (2000). On Representation Theorems for Nonmonotonic Consequence Relations. *Journal of Symbolic Logic*, *65*, 1321–1337.

Rabinowicz, W. (2008). Value Relations. *Theoria*, *74*, 18–49.

Rott, H. (1992). Preferential Belief Change Using Generalized Epistemic Entrenchment. *Journal of Logic, Language and Information*, *1*, 45–78.

Rott, H. (1993). Belief Contraction in the Context of the General Theory of Rational Choice. *Journal of Symbolic Logic*, *58*, 1426–1450.

Rott, H. (2001). *Change, Choice and Inference: A Study in Belief Revision and Nonmonotonic Reasoning*. Oxford: Oxford University Press.

Rott, H. (2003). Basic Entrenchment. *Studia Logica*, *73*, 257–280.

Rott, H. (2009). Shifting Priorities: Simple Representations for Twenty-Seven Iterated Theory Change Operators. In D. Makinson, J. Malinowski, & H. Wansing (Eds.), *Towards Mathematical Philosophy* (pp. 269–296). Berlin: Springer.

Rott, H., & Hansson, S. O. (2014). Safe Contraction Revisited. In S. O. Hansson (Ed.), *David Makinson on Classical Methods for Nonclassical Problems* (pp. 35–70). Dordrecht: Springer.

Schlechta, K. (1997). *Nonmonotonic Logics: Basic Concepts, Results, and Techniques*. Berlin: Springer.

Sen, A. K. (1986). Social Choice Theory. In K. J. Arrow & M. D. Intriligator (Eds.), *Handbook of Mathematical Economics* (Vol. 3, pp. 1073–1181). Amsterdam: Elsevier.

Suppes, P., Krantz, D. H., Luce, R. D., & Tversky, A. (1989). *Foundations of Measurement* (Vol. 2). New York: Academic Press.

Wiener, N. (1914). A Contribution to the Theory of Relative Position. *Proceedings of the Cambridge Philosophical Society*, *17*, 441–449.

Hans Rott
University of Regensburg
Germany
E-mail: hans.rott@ur.de

Relating Logics of Justifications and Evidence

Igor Sedlár[1]

Abstract: The paper relates evidence and justification logics, both philosophically and technically. On the philosophical side, it is suggested that the difference between the approaches to evidence in the two families of logics can be explained as a result of their focusing on two different notions of support provided by evidence. On the technical side, a justification logic with operators pertaining to both kinds of support is shown to be sound and complete with respect to a special class of awareness models. In addition, a realization theorem with respect to $\mathbf{K}$ is shown to hold for the logic.

Keywords: awareness logics, completeness, epistemic logic, evidence logics, justification logics, realization

1 Introduction

It is commonly assumed that normal modal epistemic logics (Fagin, Halpern, Moses, & Vardi, 1995; Hintikka, 1962; Meyer, 2001; Meyer & van der Hoek, 1995; van Benthem, 2011; van Ditmarsch, van der Hoek, & Kooi, 2008) focus on the *implicit beliefs* of an agent without being able to represent the *evidence* the agent might use to *justify* her beliefs. To represent evidence, it is argued, normal epistemic logics have to be extended. Two families of such extensions have recently risen into prominence: *justification logics* (Artemov, 2001, 2008, 2011) and *evidence logics* (Shi, 2013; van Benthem, Fernández-Duque, & Pacuit, 2012, 2014; van Benthem & Pacuit, 2011a, 2011b). Justification logics originate in provability logic and are, at least semantically, close to awareness logics of Fagin and Halpern (1988). Pieces of evidence are represented 'syntactically', as sets of formulas justified by the respective pieces, see (Artemov, 2012). Evidence logics build

---

[1]The author would like to thank Johan van Benthem for discussion and encouragement, the organizers of Logica 2013 for a pleasant conference and the editors of The Logica Yearbook for their patience. Work on this paper was carried out at the Institute of Philosophy of the Slovak Academy of Sciences as a part of the research project "Language and Determination of Meaning in Context", funded by the grant VEGA 2/0019/12.

on an evidence-based interpretation of neighbourhood models for classical modal logics, see (Chellas, 1980), and are a combination of classical and normal modal logics. Pieces of evidence are represented 'semantically', as sets of worlds consistent with the respective pieces. The difference in their respective representations of evidence makes the investigation of their relationship and combinations rather interesting.[2]

This paper takes first steps to relate these two families of logics, both philosophically and technically. Simple versions of evidence and justification logics are outlined in Sections 2 and 3, respectively. Section 4 discusses the differences between evidence and justification logics. Three prima facie differences are pointed out and explained away. First, it is shown in Section 4.1 that the basic evidence logic discussed in Section 2 is sound and complete with respect to a class of models where 'pieces of evidence' are considered explicitly. Second, it is shown in the same section that evidence logics are consistent with a 'world-relative' construal of evidence. Third, it is argued in Section 4.2 that the difference between the renderings of evidence embodied in evidence and justification logics can be explained as a result of their focusing on two different *kinds of support* provided by pieces of evidence. This is an alternative to the provisional explanations of the difference known from the literature, which tend to point out 'different levels of analysis' as the main contrast, see (van Benthem et al., 2014, pp. 108, 132), for example. As a result, combinations of evidence and justification logics are a natural research program. However, only a very simple combination is provided here: Section 5 points out that a specific version of multi-dimensional awareness logic combines ideas related to evidence and justification logics. The main technical result is a completeness theorem for this combination, together with a realization theorem with respect to the modal logic **K**. The concluding Section 6 points out that a reformulation of evidence and justification logics in the framework of *term-modal logics*, see (Fitting, Thalmann, & Voronkov, 2001), is an interesting topic for further research.

---

[2]See (van Benthem et al., 2014, p. 132), for example. An interesting combination of justification logic with dynamic epistemic logic is put forward in (Baltag, Renne, & Smets, 2012, 2014).

2 A simple evidence logic

For sake of simplicity, only the basic evidence logic without dynamic and plausibility operators is discussed. The language $\mathcal{L}_E$ adds to the Boolean language the monadic operators $[E]$, $[B]$ and $[K]$. '$[E]\phi$' is read 'there is evidence supporting ϕ' (or 'the agent has evidence for ϕ'), '$[B]\phi$' means 'the agent believes that ϕ' and the operator $[K]$ is construed as a universal modality ('$[K]\phi$' may be read as 'the agent knows that ϕ').

The simplest models for $\mathcal{L}_E$ are *extended evidence models*

$$M^e = \langle W, R, E, V \rangle \tag{1}$$

where $\langle W, R, V \rangle$ is a one-dimensional Kripke model and E is an 'evidence relation' $E \subseteq W \times 2^W$. It is required that $\langle w, W \rangle \in E$ and $\langle w, \emptyset \rangle \notin E$ for all $w \in W$. Hence, extended evidence models are a combination of Kripke models with neighbourhood models, see (Chellas, 1980; Hansen, Kupke, & Pacuit, 2009). Sets $X \in E(w)$ represent the evidence available to the agent at w (agent's 'evidential state').[3] It is assumed in addition to the usual Boolean truth-conditions that ($\|\phi\|_{M^e} = \{w : M^e, w \models \phi\}$):

- $M^e, w \models [B]\phi$ iff $R(w) \subseteq \|\phi\|_{M^e}$.
- $M^e, w \models [E]\phi$ iff there is $X \in E(w)$ such that $X \subseteq \|\phi\|_{M^e}$.
- $M^e, w \models [K]\phi$ iff $\|\phi\|_{M^e} = W$.

R is construed as an 'epistemic accessibility' relation and $[B]$ is the usual implicit belief operator. The literature on evidence logics does not offer a detailed explanation of the relation between pieces (sources) of evidence and sets $X \in E(w)$, but the following might be plausible. Let us assume that we have a set P of pieces of evidence. These might include 'propositional evidence' (i.e. statements such as 'There is a table in front of Alice') as well as non-propositional entities such as sense-experiences (Alice's visual experience of a table) etc.[4] Now assume that every $x \in P$ comes with $C(x) \subseteq W$, the set of worlds *consistent* with x. A preliminary explanation

---
[3] $E(w) = \{X : EwX\}$ and similarly for $R(w)$.

[4] It is customary in philosophy of science to assume that all evidence is propositional, see (Achinstein, 2001, 2010). On the other hand, the use of 'evidence' in epistemology is broader, including propositional as well as non-propositional entities. For nice examples, see (Feldman, 1988, 1995; Feldman & Conee, 1985), where sense-experiences are frequently cited as evidence. However, some epistemologists imply that evidence is exclusively propositional, see (Williamson, 2000), for example.

of the invoked notion of consistency might run as follows. If $x \in P$ is propositional, then $w \in C(x)$ iff x is true in w (our assumption is that w is *maximally* consistent—every statement consistent with w is true in w). If $x \in P$ is non-propositional (sense-experience, event, object etc.), then $w \in C(x)$ iff x exists (obtains) in w. Every $X \in E(w)$ can be seen as corresponding to some $x \in P$ in the sense that $X = C(x)$. The set $E(w)$ then corresponds to $P(w) \subseteq P$, the evidence available to the agent at w. $[E]\phi$ then holds in w iff there is a piece of evidence x such that x is available at w and ϕ holds in every world in which x holds (or exists, obtains, occurs etc.), i.e. x 'necessitates' ϕ.[5]

Definition 1 (van Benthem et al., 2012) *The Hilbert system $H(\mathbf{EL})$ is given by the following axiom schemes and rules:*

(A0) Propositional tautologies in $\mathcal{L}_E$,

(A1) **S5** *axioms for $[K]$,*

(A2) **K** *axioms for $[B]$,*

(A3) $[E]\top$,

(A4) $([E]\phi \wedge [K]\psi) \leftrightarrow [E](\phi \wedge [K]\psi)$,

(A5) $[K]\phi \to [B][K]\phi$,

(R1) Modus Ponens,

(R2) $\phi \to \psi / [E]\phi \to [E]\psi$,

(R3) $\phi / [o]\phi$ for $o \in \{K, B\}$.

The basic evidence logic **EL** *is the set of formulas provable in $H(\mathbf{EL})$.*

Fact 1 (van Benthem et al., 2012) *For every $\phi \in \mathcal{L}_E$: $\phi \in$* **EL** *iff $M^e, w \models \phi$ for all pointed extended evidence models M^e, w.*

---

[5]This is in need of a deeper discussion. However, such a discussion is left out of the present paper, due to space limitations. We note that our preliminary characterisation of consistency loosely builds on Feldman's characterisation of the relation of 'necessitation' between pieces of evidence and propositions, see (Feldman, 1995).

3 A simple justification logic

The basic language of justification logic $\mathcal{L}_J$ adds to the Boolean language formulas of form $t : \phi$, where $t \in Tm$. The set of *justification terms* Tm is defined inductively over disjoint countable sets Var of justification variables and Con of justification constants (x, y, z range over Var and d, e range over Con):

- $Var \cup Con \subseteq Tm$.
- If $s, t \in Tm$, then $s \cdot t \in Tm$ and $s + t \in Tm$.

Hence, in addition to 'tags' for specific pieces of evidence (or justifications), $\mathcal{L}_J$ contains *operators*, which allow to build complex justification terms. Formulas $t : \phi$ are read 'ϕ is believed for reason t'.

Definition 2 *The Hilbert system $H(\mathbf{J})$ comprises of the following axioms and rules:*

(jA0) Propositional tautologies,

(jA1) $s : (\phi \to \psi) \to (t : \phi \to (s \cdot t) : \psi)$,

(jA2) $(s : \phi \vee t : \phi) \to (s + t) : \phi$,

(jR1) Modus Ponens,

(jR2) For every axiom ϕ and any constants $e_1, \ldots, e_n$ infer that $e_n : e_{n-1} : \ldots : e_1 : \phi$.

The constant specification induced by $H(\mathbf{J})$, $CS_\mathbf{J}$, is the set of all formulas of the form $e : \phi$, where e is a constant, provable in $H(\mathbf{J})$. The basic justification logic $\mathbf{J}$ is the set of $\mathcal{L}_J$-formulas provable in $H(\mathbf{J})$.

The justification logic $\mathbf{J}$ has two interesting properties: it 'internalizes' its own proofs and it 'realizes' the basic normal modal logic $\mathbf{K}$.

Theorem 1 (Internalization; Brezhnev, 2000) *If $\phi \in \mathbf{J}$, then there is a t such that $t : \phi \in \mathbf{J}$.*

Theorem 2 (K-realization; Brezhnev, 2000)
  i) *If $\phi \in \mathbf{K}$ then there is a formula $\phi^r \in \mathcal{L}_J$ such that ϕ^r results from ϕ by replacing occurrences of 'boxes' by justification operators and $\phi^r \in \mathbf{J}$.*
  ii) *If $\phi \in \mathbf{J}$ and $\phi^\square$ results from ϕ by replacing every occurrence of a justification operator by a 'box', then $\phi^\square$ is a theorem of $\mathbf{K}$.*

The usual models for $\mathcal{L}_J$ are *Fitting models*, see (Fitting, 2005):

$$M^f = \langle W, R, A, V \rangle \qquad (2)$$

where $\langle W, R, V \rangle$ is a one-dimensional Kripke model and A is a function from $(W \times Tm)$ to 2^{Fm} such that:

- If $\phi \to \psi \in A(w, s)$ and $\phi \in A(w, t)$, then $\psi \in A(w, s \cdot t)$.
- $A(w, s) \cup A(w, t) \subseteq A(w, s + t)$.
- If $e : \phi \in CS_\mathbf{J}$, then $\phi \in A(w, e)$ for all $w \in W$.

The truth-conditions of the Boolean fragment are as usual. Moreover:

- $M^f, w \models s : \phi$ iff i) $M^f, v \models \phi$ for all v such that Rwv and ii) $\phi \in A(w, s)$.

The relation R is construed in the usual way. The set $A(w, s)$ is seen as the set of formulas justified by s at w. This is a 'syntactic filter' akin to the awareness function of Fagin and Halpern (1988) with an extra parameter, the justification term.[6] Note that 't justifies ϕ' is world-relative: typically $A(w, t) \neq A(v, t)$ for $w \neq v$.

The operator '$\cdot$' corresponds to applying Modus Ponens (it is often called 'application'): If s justifies $\phi \to \psi$ and t justifies ϕ, then $s \cdot t$ justifies ψ. No other properties of '$\cdot$' are assumed.[7] The operator '$+$' ('sum' or 'weakening') corresponds to 'monotonic merging' of justifications: If $\phi \in A(w, s)$ or $\phi \in A(w, t)$, then $\phi \in A(w, s + t)$.[8] Formula $t : \phi$ holds at w iff the agent implicitly believes that ϕ and t justifies ϕ at w. Hence, the precise meaning of $t : \phi$ could be spelled out as 'agent's implicit belief that ϕ is justifiable by reference to t'. Note that justification logics do not work with the notion of some justifications being 'available'.[9]

Fact 2 (Fitting, 2005) $\phi \in \mathbf{J}$ *iff ϕ is valid in every Fitting model.*

---

[6]Of course, the function A can be replaced by a family of awareness functions $\{A_s\}_{s \in Tm}$. We will return to this suggestion later.

[7]In particular, '$\cdot$' is not assumed to be commutative (there are models with $A(w, s \cdot t) \neq A(w, t \cdot s)$), associative ($A(w, s \cdot (t \cdot t')) \neq A(w, (s \cdot t) \cdot t')$), nor idempotent ($A(w, s) \neq A(w, s \cdot s)$). Applying '$\cdot$' is 'non-monotonic', as it may lead to 'forgetting': there are models with $A(w, s) \not\subseteq A(w, s \cdot t)$.

[8]Commutativity, associativity and idempotence are not assumed, although $A(w, t) \subseteq A(w, t + t)$. 'Forgetting' is ruled out.

[9]However, availability of t at w could be mimicked (at least in the mono-agent case) by $A(w, t) \neq \emptyset$.

4 What is the difference?

A preliminary characterisation of the difference between evidence and justification logics is now at hand. There are at least three interesting points:

1. *Reference to pieces of evidence.* Evidence logics do not refer to specific pieces of evidence explicitly, 'syntactically' ($\mathcal{L}_E$) nor 'semantically' (extended evidence models). On the other hand, justification logics refer to 'justifications' both syntactically (justification terms in $\mathcal{L}_J$) and semantically (A in Fitting models).

2. *What does it mean to have evidence for ϕ?* In extended evidence models, 'there is evidence for ϕ' is a function of the proposition expressed by ϕ. On the other hand, the function A in Fitting models picks formulas directly, without reference to the propositions expressed.

3. *Relativity of justifications.* In Fitting models, 't justifies ϕ' is world-relative. On the other hand, it is not clear if this is the case in extended evidence models. Our explanation in Section 2 suggests that extended evidence models are consistent with a 'constant-evidence' explanation.

We show in Section 4.1 that direct reference to pieces of evidence, at least on the 'level of models', is easily added to evidence logics. However, adding 'tags' for pieces of evidence to $\mathcal{L}_E$ is more complicated and we leave it for another occasion. It is also shown that evidence logics are consistent with a 'world-relative' construal of evidence. Item 2 makes it tempting to conclude that justification logics 'go deeper', beyond the semantic level of propositions. Section 4.2 offers a different explanation, according to which the difference results from focusing on two different kinds of support.

4.1 Evidence logics and pieces of evidence

Extended evidence models can be 'safely' replaced by models that *do* invoke specific pieces of evidence. Neighbourhoods can be simulated by sets of binary relations.

Definition 3 *A two-sorted evidence model is a tuple*

$$\mathcal{M} = \langle W, R, \{R_i\}, S, V \rangle_{i \in G} \tag{3}$$

where $\langle W, R, \{R_i\}, V\rangle_{i \in G}$ is a $(|G|+1)$-dimensional Kripke model and S: $W \to 2^G$. The truth-conditions for Boolean connectives, $[B]\phi$ and $[K]\phi$ are as in extended evidence models. Moreover:

- $\mathcal{M}, w \models [E]\phi$ iff there is $i \in S(w)$ such that $R_i(w) \subseteq \|\phi\|_{\mathcal{M}}$.

A two-sorted model is *extendible iff (i) for every w, there is $i \in S(w)$ such that $R_i(w) = W$, and (ii) if $i \in S(w)$, then $R_i(w) \neq \emptyset$.*

The set G is thought of as a set of pieces of evidence and every R_i is a binary relation of 'relative compatibility' corresponding to $i \in G$. $R_i wv$ can be thought of as representing the fact that v is compatible with i, relatively to w. $S(w)$ is seen as the body of evidence available to the agent at w. Sets $R(w)$ can be seen as corresponding to the evidence the agent 'accepts' or 'trusts' at w. One can think of $R(w)$ as the intersection of a multitude of sets corresponding to specific 'accepted' pieces of evidence, but we shall not go into such details here. The models recognize two sorts of evidence, hence their name. Observe that no specific relation between the 'available' and the 'accepted' evidence is assumed.

Definition 4 *Let $\mathcal{M} = \langle W, R, \{R_i\}, S, V\rangle_{i \in G}$ be a two-sorted evidence model. The* extended copy *of $\mathcal{M}$, $\mathcal{M}^* = \langle W, R, E^*, V\rangle$, is obtained by defining:*

- $E^*(w) = \{X : X = R_i(w) \text{ for some } i \in S(w)\}$.

Fact 3 *Let $\mathcal{M}$ be an extendible two-sorted model. Then $\mathcal{M}^*$ is an extended evidence model. Moreover, if M^e is an arbitrary extended evidence model, then there is an extendible two-sorted model $\mathcal{M}$ such that $M^e = \mathcal{M}^*$.*

Fact 4 *Let $M^e = \mathcal{M}^*$. Then $\|\phi\|_{M^e} = \|\phi\|_{\mathcal{M}}$ for every $\phi \in \mathcal{L}_E$.*

Theorem 3 *For all $\phi \in \mathcal{L}_E$: $\phi \in$ **EL** iff $\mathcal{M}, w \models \phi$ for all pointed extendible two-sorted models $\mathcal{M}, w$.*

Proof. Soundness is easily established by induction on the length of proofs. Completeness follows from Facts 1, 3 and 4. □

Note that the evidence-based interpretation of two-sorted models extends to simple multi-dimensional Kripke models $M = \langle W, \{R_i\}, V\rangle_{i \in G}$. In other words, normal multi-dimensional modal logics can be construed as simple logics of evidence, where 'agents' are replaced by 'pieces of evidence' and 'groups' by 'evidential states'.

4.2 Two kinds of support

Evidence logics are based on the notion that ϕ is supported by evidence iff there is an available piece of evidence that *necessitates* ϕ, i.e. every world in which the piece of evidence holds (or obtains) is a ϕ-world (call this the *propositional* notion of support). Consequently, if ϕ and ψ express the same proposition, then ϕ is propositionally supported iff ψ is. It is plain that this is not the case in the context of justification logics, where ϕ is supported by evidence (at w) iff $\phi \in A(w,t)$ for some t. It is possible that ϕ, ψ express the same proposition in a Fitting model and $\phi \in A(w,t)$, but $\psi \notin A(w,s)$ for all s.

The propositional notion of support is assumed by many sceptical arguments. A typical sceptical argument insists that a class of beliefs is not supported *in that* it is not necessitated by any evidence. For example, the sceptic argues that my belief that there is a table in front of me is not supported by my visual experiences, because it is possible for me to have the experiences in worlds where there is no table in front of me. For example, it is possible to see a table while hallucinating, while being a brain in a vat etc. These possibilities are not *ruled out* by our evidence.

Accordingly, the propositional notion of support seems unrealistic from an intuitive point of view. For example, when considering perceptual beliefs, one tends to consider visual experiences as evidence *par excellence*. There is a difference between our 'intuitive' understanding of evidence and the propositional notion of support. Some epistemologists make a similar point. For example, Feldman (1995) argues that the question whether one's belief in ϕ is *rational* is independent of the question whether one's evidence necessitates ϕ.

Example 1 Let $\top$ be any propositional tautology, let p be short for 'There is a table in front of Alice' and let x denote Alice's visual experience of a table in front of her. Intuitively, x supports p and x does not support $\top$. But, it is plain that x necessitates $\top$, for the sole reason that $\top$ is necessary. Moreover, x does not necessitate p, as the obvious sceptical counterexamples show.

One can explain the difference between evidence and justification logics in terms of the contrast between the propositional and the 'intuitive' notion of support. Evidence logics represent propositional support and justification logics focus on the independent intuitive notion. On this interpretation of the difference, one immediately sees the relevance of logics that *combine* these two approaches.

5 A simple two-sorted justification logic

A simple logic that represents the propositional as well as the 'intuitive' notion of support is introduced in the present section (5.1). The logic is shown to be sound and complete with respect to a class of multi-dimensional awareness models (5.2). In addition, it is shown that the logic realizes **K**, but the usual proofs of internalization fail (5.3).

5.1 The logic JE

The language $\mathcal{L}_{JE}$ is $\mathcal{L}_J$ extended with monadic operators $[E_t]$ and $[A_t]$ for every $t \in Tm$. $[E_t]\phi$ is read 't necessitates ϕ' and $[A_t]\phi$ is read 't weakly supports ϕ'. The justification formulas $t : \phi$ are read 't strongly supports ϕ'. Strong support is construed as a combination of necessitation (propositional support) and weak ('intuitive') support.

Definition 5 *The Hilbert system $H(\mathbf{JE})$ is given by the following axioms and rules:*

(Ax0) Propositional tautologies,

(Ax1) $[E_t](\phi \to \psi) \to ([E_t]\phi \to [E_t]\psi)$,

(Ax2) $([E_s]\phi \vee [E_t]\phi) \to ([E_{s \cdot t}]\phi \wedge [E_{s+t}]\phi)$,

(Ax3) $([E_s]\phi \wedge [E_t]\psi) \to ([E_{s \cdot t}](\phi \wedge \psi) \wedge [E_{s+t}](\phi \wedge \psi))$,

(Ax4) $[A_s](\phi \to \psi) \to ([A_t]\phi \to [A_{s \cdot t}]\psi)$,

(Ax5) $([A_s]\phi \vee [A_t]\phi) \to [A_{s+t}]\phi$,

(Ax6) $t : \phi \leftrightarrow ([E_t]\phi \wedge [A_t]\phi)$,

(Ax7) $s : (\phi \to \psi) \to (t : \phi \to (s \cdot t) : \psi)$,

(Ax8) $(s : \phi \vee t : \phi) \to (s + t) : \phi$,

(Ru1) Modus Ponens,

(Ru2) $\phi/[E_t]\phi$ *for all* $t \in Tm$,

(Ru3) For every axiom ϕ and any constants $e_1, \ldots, e_n$ infer that $e_n : e_{n-1} : \ldots : e_1 : \phi$.

Logics of Justifications and Evidence 217

The constant specification induced by $H(\mathbf{JE})$, $CS_{\mathbf{JE}}$, *is the set of formulas of the form* $e : \phi$ *provable in* $H(\mathbf{JE})$. *The logic* $\mathbf{JE}$ *is the set of formulas provable in* $H(\mathbf{JE})$.

Lemma 1 $\mathbf{J} \subseteq \mathbf{JE}$.

Lemma 2 *Axioms (Ax7) and (Ax8) are redundant, i.e. derivable from the other axioms.*

We note that Lemma 1 requires the inclusion of the 'redundant' axioms (Ax7) and (Ax8), since (Ru3) applies only to axioms. The constant specification induced by the Hilbert system without the two axioms does not contain $CS_{\mathbf{J}}$.

Definition 6 *A common model is a tuple*

$$\mathfrak{M} = \langle W, \{R_t\}, \{A_t\}, V \rangle_{t \in Tm} \qquad (4)$$

where every $R_t \subseteq W^2$ *and* $A_t : W \to 2^{Fm(\mathcal{L}_{JE})}$. *It is assumed that*

- $R_{s \cdot t}, R_{s+t} \subseteq R_s \cap R_t$.
- *If* $\phi \to \psi \in A_s(w)$ *and* $\phi \in A_t(w)$, *then* $\psi \in A_{s \cdot t}(w)$.
- $A_s(w) \cup A_t(w) \subseteq A_{s+t}(w)$.
- *If* $e : \phi \in CS_{\mathbf{JE}}$, *then* $\phi \in A_e(w)$ *for all* $w \in W$.

The truth-conditions for the Boolean fragment are as usual. Moreover:

$\mathfrak{M}, w \models [E_t]\phi$ *iff* $R_t(w) \subseteq \|\phi\|_\mathfrak{M}$. $\quad(*)$

$\mathfrak{M}, w \models [A_t]\phi$ *iff* $\phi \in A_t(w)$. $\quad(**)$

$\mathfrak{M}, w \models t : \phi$ *iff* $(*)$ *and* $(**)$.

Common frames *and* validity *are defined in the usual way.*

Common models are a special class of multi-dimensional awareness models. For every $t \in Tm$, R_t represents the propositional support provided by t and A_t represents the weak ('intuitive') support. $R_t(w)$ is the set of worlds consistent with t relatively to w. $A_t(w)$ is the set of formulas weakly supported by t at w. Note that these are independent: there are models where $R_t(w) \subseteq \|\phi\|_\mathfrak{M}$ but $\phi \notin A_t(w)$ and vice versa. **EJ** can be seen as a justification logic that incorporates the notion of propositional support from evidence logic. However, the logic does not contain **EL** for the simple reason that it replaces the idea of 'quantifying' over pieces of evidence by operators expressing the propositional support provided by *specific* pieces of evidence.

5.2 Completeness

This section establishes the usual soundness and completeness results. Completeness is shown by the standard canonical model construction. We note that the strong justification logic **LP** was shown to be sound and complete with respect to a special class of common models in (Sedlár, 2013).

Theorem 4 (Soundness) *If $\phi \in$ **JE**, then ϕ is valid in every common frame.*

Proof. Induction on the length of proofs. □

Definition 7 (Canonical frame and model) *The* canonical frame *for* **JE** *is a structure* $\mathfrak{F}^c = \langle W^c, \{R_t^c\}, \{A_t^c\}\rangle_{t \in Tm}$ *where*

- W^c *is the set of maximal* **JE**-*consistent sets of formulas* $\Gamma, \Delta, \ldots$
- $\Gamma R_t^c \Delta$ *iff* $\Gamma_E(t) \subseteq \Delta$, *where* $\Gamma_E(t) = \{\phi : [E_t]\phi \in \Gamma\}$.
- $\phi \in A_t^c(\Gamma)$ *iff* $[A_t]\phi \in \Gamma$.

The canonical model *for* **JE** *is* $\mathfrak{M}^c = \langle \mathfrak{F}^c, V^c \rangle$, *where* $\Gamma \in V^c(p)$ *iff* $p \in \Gamma$.

Lemma 3 (Frame Lemma) *The canonical frame is a common frame.*

Proof. As usual, we have to show that the canonical frame satisfies the frame conditions of Definition 6. First, it has to be shown that $\Gamma R_s^c \Delta$ and $\Gamma R_t^c \Delta$ if $\Gamma R_{s+t}^c \Delta$. If $\phi \in \Gamma_E(s)$, then $[E_s]\phi \in \Gamma$ and, by (Ax2) and propositional logic, $[E_{s+t}]\phi \in \Gamma$. By the assumption, $\phi \in \Delta$. The cases for $\Gamma R_t^c \Delta$ and $R_{s \cdot t}^c$ are similar.

Second, assume that $\phi \to \psi \in A_s^c(\Gamma)$ and $\phi \in A_t^c(\Gamma)$. $\psi \in A_t^c(\Gamma)$ follows from (Ax4). The fact that $A_s^c(\Gamma) \cup A_t^c(\Gamma) \subseteq A_{s+t}^c(\Gamma)$ is proven similarly by invoking (Ax5).

Third, assume that $e : \phi \in CS_{\mathbf{JE}}$. Then $e : \phi \in$ **JE** and $e : \phi \in \Gamma$ for all $\Gamma \in W^c$. By (Ax6) and propositional logic, $[A_e]\phi \in \Gamma$ for all $\Gamma \in W^c$. Hence, $\phi \in A_e^c(\Gamma)$ for all Γ. □

Lemma 4 (Model Lemma) *The canonical model is a common model.*

Proof. The Lemma follows from Lemma 3 and the fact that $\phi \in \Gamma$ iff $\mathfrak{M}^c, \Gamma \models \phi$. The second claim (Truth Lemma) is shown by standard induction on the complexity of ϕ. The base case holds by definition and the cases for Boolean connectives are trivial.

Now assume that $[E_t]\phi \in \Gamma$. It follows that $\phi \in \Gamma_E(t)$ and, hence, $\phi \in \Delta$ for all $\Gamma_E(t) \subseteq \Delta$. Consequently, $\mathfrak{M}^c, \Gamma \models [E_t]\phi$. Conversely, assume that $[E_t]\phi \notin \Gamma$. Then $\Gamma_E(t) \cup \{\neg\phi\}$ is **JE**-consistent and can be extended to a maximal **JE**-consistent set Γ^*. It is plain that $\Gamma R_t^c \Gamma^*$ and $\phi \notin \Gamma^*$. Hence, $\mathfrak{M}^c, \Gamma \not\models [E_t]\phi$.

Next, $[A_t]\phi \in \Gamma$ iff $\phi \in A_t^c(\Gamma)$ (by definition) iff $\mathfrak{M}^c, \Gamma \models [A_t]\phi$. Now $t : \phi \in \Gamma$ iff $[E_t]\phi \in \Gamma$ and $[A_t]\phi \in \Gamma$ by (Ax6). The rest follows from the previous claims concerning $[E_t]\phi$ and $[A_t]\phi$. □

Theorem 5 (Completeness) *If ϕ is valid in every common frame, $\phi \in$ **JE**.*

5.3 Realization and internalization

In this section, an 'operator' is any instance of '$[E_t]$', '$[A_t]$' and '$t :$', and a 'justification operator' is any instance of '$t :$'.

Lemma 5 *Let $\phi \in \mathcal{L}_{JE}$ and let $\phi^\square$ be the result of replacing every occurrence of an operator in ϕ by an occurrence of the modal box '$\square$'. If $\phi \in$ **JE**, then $\phi \in$ **K**.*

Proof. Simple induction on the length of $H(\mathbf{JE})$-proofs. Observe that the claim holds for every axiom of $H(\mathbf{JE})$ and the rules 'preserve the claim' as well. □

Theorem 6 *If $\phi \in$ **K** then there is a formula $\phi^r \in \mathcal{L}_{JE}$ such that ϕ^r results from ϕ by replacing occurrences of 'boxes' by occurrences of justification operators and $\phi^r \in$ **JE**.*

Proof. Follows from Theorem 2 and Lemma 1. □

A corollary of these two results is that **JE** 'realizes' **K**. However, the usual proof of the internalization property (see, e.g., Artemov, 2008) does not work. The reason is that there is no justification operator corresponding to the necessitation rule (Ru2). Moreover, other well-known techniques used when justification logic is combined with normal modal logics (see Artemov & Nogina, 2005a, 2005b for example) are not applicable in our context either.

6 Conclusion

The paper attempted to take first steps to relate evidence and justification logics. The main results are: (i) a completeness proof for the basic evidence logic with respect to a new class of models, where 'pieces of evidence' are invoked explicitly, (ii) completeness and realization proofs for a justification logic that incorporates some ideas form evidence logic (operators for propositional support). In a more philosophical vein, it has been suggested that (i) the difference between the rendering of evidence in justification and evidence logics can be explained as the result of their focusing on two distinct notion of support, (ii) even multi-dimensional normal modal logics can be seen as logics of evidence.

However, a natural goal is to extend **JE** at least with the operator $[E]$ of evidence logic. Of course, this could be done by adding neighbourhoods to common models. A more interesting approach is to construe $[E]$ as a quantifier over selected subsets of justification terms. Such a framework comes close to term-modal logics of Fitting et al. (2001), a version of first-order modal logic where modal operators are indexed by the terms of the language. This approach makes the introduction of predicates for and quantification over pieces of evidence relatively straightforward. However, this interesting project is beyond the scope of this paper.

References

Achinstein, P. (2001). *The Book of Evidence*. Oxford: Oxford University Press.

Achinstein, P. (2010). Concepts of Evidence. In *Evidence, Explanation, and Realism. Essays in Philosophy of Science* (pp. 3–33). Oxford: Oxford University Press.

Artemov, S. (2001). Explicit Provability and Constructive Semantics. *Bulletin of Symbolic Logic, 7*, 1–36.

Artemov, S. (2008). The Logic of Justification. *The Review of Symbolic Logic, 1*, 477–513.

Artemov, S. (2011). Why Do We Need Justification Logic? In J. van Benthem, A. Gupta, & E. Pacuit (Eds.), *Games, Norms and Reasons: Logic at the Crossroads* (pp. 23–38). Dordrecht: Springer.

Artemov, S. (2012). The Ontology of Justifications in the Logical Setting. *Studia Logica, 100*, 17–30.

Artemov, S., & Nogina, E. (2005a). Introducing Justification into Epistemic Logic. *Journal of Logic and Computation, 15,* 1059–1073.

Artemov, S., & Nogina, E. (2005b). On Epistemic Logic with Justification. In R. van der Meyden (Ed.), *Theoretical Aspects of Rationality and Knowledge, Proceedings of the Tenth Conference (TARK 2005)* (pp. 279–294). Singapore: National University of Singapore.

Baltag, A., Renne, B., & Smets, S. (2012). The Logic of Justified Belief Change, Soft Evidence and Defeasible Knowledge. In L. Ong & R. de Queiroz (Eds.), *Proceedings of the 19th Workshop on Logic, Language, Information and Computation (WoLLIC 2012)* (pp. 168–190). Berlin, Heidelberg: Springer.

Baltag, A., Renne, B., & Smets, S. (2014). The Logic of Justified Belief, Explicit Knowledge, and Conclusive Evidence. *Annals of Pure and Applied Logic, 165,* 49–81.

Brezhnev, V. N. (2000). *On Explicit Counterparts of Modal Logics* (Tech. Rep. Nos. CFIS 2000–05). Ithaca: Cornell University.

Chellas, B. (1980). *Modal Logic: An Introduction.* Cambridge: Cambridge University Press.

Fagin, R., & Halpern, J. Y. (1988). Belief, Awareness, and Limited Reasoning. *Artificial Intelligence, 34,* 39–76.

Fagin, R., Halpern, J. Y., Moses, Y., & Vardi, M. Y. (1995). *Reasoning About Knowledge.* Cambridge, MA: MIT Press.

Feldman, R. (1988). Having Evidence. In D. Austin (Ed.), *Philosophical Analysis* (pp. 83–104). Dordrecht: Kluwer Academic Publishers.

Feldman, R. (1995). Authoritarian Epistemology. *Philosophical Topics, 23,* 147–169.

Feldman, R., & Conee, E. (1985). Evidentialism. *Philosophical Studies, 48,* 15–34.

Fitting, M. (2005). The Logic of Proofs, Semantically. *Annals of Pure and Applied Logic, 132,* 1–25.

Fitting, M., Thalmann, L., & Voronkov, A. (2001). Term-modal Logics. *Studia Logica, 69,* 133–169.

Hansen, H. H., Kupke, C., & Pacuit, E. (2009). Neighbourhood Structures: Bisimilarity and Basic Model Theory. *Logical Methods in Computer Science, 5,* 1–38.

Hintikka, J. (1962). *Knowledge and Belief.* Ithaca: Cornell University Press.

Meyer, J.-J. C. (2001). Epistemic Logic. In L. Goble (Ed.), *The Blackwell Guide to Philosophical Logic* (pp. 183–202). Oxford: Blackwell.

Meyer, J.-J. C., & van der Hoek, W. (1995). *Epistemic Logic for AI and Computer Science*. Cambridge: Cambridge University Press.

Sedlár, I. (2013). Justifications, Awareness and Epistemic Dynamics. In S. Artemov & A. Nerode (Eds.), *Logical Foundations of Computer Science 2013 (LNCS 7734)* (pp. 307–318). Berlin, Heidelberg: Springer.

Shi, C. (2013). Logic of Evidence-based Knowledge. In D. Grossi, O. Roy, & H. Huang (Eds.), *Logic, Rationality, and Interaction (LNCS 8196)* (pp. 347–351). Berlin, Heidelberg: Springer.

van Benthem, J. (2011). *Logical Dynamics of Information and Interaction*. Cambridge: Cambridge University Press.

van Benthem, J., Fernández-Duque, D., & Pacuit, E. (2012). Evidence Logic: A New Look at Neighborhood Structures. In T. Bolander, T. Braüner, S. Ghilardi, & L. Moss (Eds.), *Advances in Modal Logic 9* (pp. 97–118). London: College Publications.

van Benthem, J., Fernández-Duque, D., & Pacuit, E. (2014). Evidence and Plausibility in Neighborhood Structures. *Annals of Pure and Applied Logic*, *165*, 106–133.

van Benthem, J., & Pacuit, E. (2011a). Dynamic Logics of Evidence-based Beliefs. *Studia Logica*, *99*, 61–92.

van Benthem, J., & Pacuit, E. (2011b). Logical Dynamics of Evidence. In H. van Ditmarsch, J. Lang, & S. Ju (Eds.), *Logic, Rationality, and Interaction (LNCS 6953)* (pp. 1–27). Berlin, Heidelberg: Springer.

van Ditmarsch, H., van der Hoek, W., & Kooi, B. (2008). *Dynamic Epistemic Logic*. Dordrecht: Springer.

Williamson, T. (2000). *Knowledge and Its Limits*. Oxford: Oxford University Press.

Igor Sedlár
Institute of Philosophy, Slovak Academy of Sciences
Slovakia
E-mail: igor.sedlar@savba.sk

Erotetic Epistemic Logic in Private Communication Protocol

PETR ŠVARNÝ[1], ONDREJ MAJER[2]
AND MICHAL PELIŠ[3]

Abstract:
The Russian Cards Problem is a toy model of safe communication via an open channel. It has been widely discussed in the literature, some of the recent approaches employ the apparatus of dynamic epistemic logic and represent communication of players by public announcements. In this article we propose a solution which adds questions to players communication toolkit. We compare it to the solutions using public announcements and provide some complexity bounds.

Keywords: russian cards problem, epistemic logic, erotetic logic, public announcement

1 Introduction

The Russian Cards Problem (RCP) is a coordination game between two agents with a third agent trying to eavesdrop on their exchange. The goal of cooperating agents is to publicly communicate each other's hand without providing any information to the third agent. The RCP problem was originally formulated on Russian Mathematics Olympiad at 2000 as a problem in information theory, but later it was introduced in the epistemic logic community by van Ditmarsch (2003). Together with his coauthors, he later presented mainly combinatorial solutions to the problem in (van Ditmarsch, van der Hoek, van der Meyden, & al., 2006). His original paper was, however, based on epistemic logic. A possible limitation of RCP is in its strict algorithm. Follow-up works to the original articles do explore the possible combinations of agents and cards involved in an RCP protocol, see (Albert,

---

[1]The work on this paper was supported by the Internal grant of the Faculty of Arts, Charles University in Prague, VG176.
[2]The work on this paper was supported by the grant GA13-21076S of the Grant Agency of the Czech Republic.
[3]The work on this paper was supported by Program for Development of Sciences at the Charles University in Prague no. 13 (Prvouk) Rationality in the Human Sciences, section Methods and Applications of Modern Logic.

Aldred, Atkinson, van Ditmarsch, & Handley, 2005; Duan & Yang, 2009). All these suggestions maintain the possibility of publicly sharing information without allowing an eavesdropper to cause harm. We will discuss the protocol in more detail later.

Nevertheless, there was no qualitative shift in the protocol. Our attempt in this article is to elaborate on the logical part of the protocol by adding tools from erotetic logic in the sense of (Peliš & Majer, 2011).

We have a few assumptions on which our work is based. First of all, agents are sincere. Hence they do not attempt any deception. They can only present true information. However, unless they are directly asked, they can choose to divulge only a part of the information available to them or add superfluous information.

Second, the agents are simple software agents. We do not assume that the agents are human-like. Hence, they do not have very complicated states and their information is stored in a well-defined system of finite statements. They also lack complex agendas to support their question posing strategies.

Last, we assume that the agents' knowledge can be represented by means of epistemic logic.

These assumptions are based on the view that a great deal of autonomous communication (for example on the Internet) can be performed by simple agents which need to deal with complex epistemic situations. An exemplar task can be the identification of a string like 20010db8142857ab. Questions enriched RCP agents could communicate such an information privately via public channels and discern, whether it is a Hexadecimal MAC address (20-01-0d-b8-14-28-57-ab), an IPv4 address (2001.0db8.1428.57ab), or an abbreviated IPv6 address (2001:0db8:0000:0000:0000:0000:1428:57ab). We will see now what obstacles we have to overcome in order to get this desired result and what we learn about questions when we use them for this purpose.

2 Russian Cards Problem

There are three agents named Anne, Bill, and Crow, abbreviated A, B, and C. Each of them receives some cards from a given stack of cards, which is known to all of them. In the case of the archetypal RCP, it is seven cards which are marked by numbers from 0 to 6. Anne and Bill get each three cards and Crow gets the last card. The state of the game can be represented as a pointed modal model over a propositional language extended with a knowledge operator K_i, common knowledge operator C, and a public announcement operator $[\psi]$. What we will call archetypal RCP can be summed up in the following way.

- Players: Anne, Bill, Crow
- Card deals (in the order A, B, C) of the type: 3 | 3 | 1 (if we want to display the actual card deal explicitly, then, e.g., 012 | 345 | 6)
- Basic goals of the problem:
    - B must be able to infer the actual hand of A (and vice versa)
    - C must not be able to infer any of A's cards
    - C must not be able to infer any of B's cards
- Common knowledge among agents: All of here mentioned except the actual deal
- Tools: Pointed (modal) models over a propositional language containing $K_i\varphi$ (individual knowledge of an agent i), $C\varphi$ (common knowledge of the group of all agents), $[\psi]\varphi$ (after publicly announcing ψ the formula φ is valid)

We should mention, that a deal 012 | 345 | 6 does not have to mean that the agents get exactly these cards. It signifies that the agents get all distinct cards and no card is repeated in the distribution. The numbers merely signify the type and are given starting from the first card of the first player. The given solutions of RCP do not depend on a particular card deal.

An example of the resulting pointed modal model is a Hexa model. This model is much simpler than a general RCP case, as it represents three agents sharing three cards. Each agent has only one card. The number sequence then shows how agents have their cards distributed. The card distribution 012 means that Anne holds the card 0, Bill the card 1 and Crow card 2. The formula describing this state is therefore:

$$012 \equiv 0_a \wedge \neg 1_a \wedge \neg 2_a \wedge \neg 0_b \wedge 1_b \wedge \neg 2_b \wedge \neg 0_c \wedge \neg 1_c \wedge 2_c \qquad (1)$$

The following Figure 1 shows three agents, each holding one card and not knowing about the distribution of the other cards.

Notice that each state represents a card deal and hence the main problem for the agents is to distinguish between card deals, i.e., states. Examples of

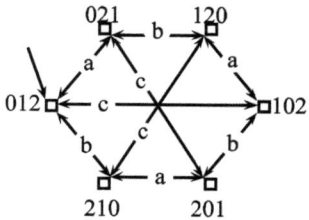

Figure 1: Hexa example from (van Ditmarsch, 2003)

simple formulae satisfied in this model are $K_a 0_a$(Anne knows she holds the card 0) or $K_b \neg K_a 1_b$ (Bill knows Anne does not know he holds the card 1).

The question that interested Albert et al. (2005) was what the conditions for *good announcements* are. A good announcement is such an announcement that the eavesdropper does not have a possibility to guess the cooperating agents' cards, but it helps the cooperating agents to advance in their goal. This is the main research question in RCP—for what card distributions are we able to get good announcements. Most of the cited works list different conditions on the numbers of cards for these safe distributions, i.e. those where a good announcement can be made. The main problem is that some of the possible card distributions do not allow for good announcements. Agents simply do not have enough cards to create an announcement that would satisfy these conditions.

In the case of an archetypal RCP, these are two examples of possible good announcements for the agent A:

- $[K_a(012_a \vee 034_a \vee 056_a \vee 135_a \vee 146_a \vee 236_a \vee 245_a)]$
- $[K_a(012_a \vee 034_a \vee 056_a \vee 135_a \vee 246_a)]$

The study in (Albert et al., 2005) lists conditions for announcements and also the card distributions that make good announcements possible. If these conditions are not met, there is no announcement which would be safe from an eavesdropper.

Already van Ditmarsch (2003) addressed the question how to deal with the security aspect of the RCP protocol and presented the following example of its application.

Example 1 There are seven cards $0, 1, \ldots, 6$. Anne's hand is 125. She knows only her own cards now, that there are seven cards, and that either Bill or Crow hold the three cards 346. She wants to find out who holds the three cards. She realizes that both 251 and 643 are prime numbers. She can now either announce: "Who is the first to tell me the factorization of 161393?" which we may expect Bill to do faster than Crow, as Bill can simply divide that number by 634 (his 'private key', so to speak), or she may announce one of the direct exchanges for hand 125, e.g.: "My hand is one of $\{125, 023, 246, 045, 356, 016, 134\}$" after which only Bill, who actually holds 346, and not Crow, is able to tell her that she holds 125. In the first case, Crow is (presumably) not fast enough ('too complex') to pose as Bill with certainty, in the second case, it is impossible to pose as Bill with certainty.

This leads to the motivation to investigate the interaction of larger numbers of agents in later works, see (Duan & Yang, 2009). This is done still along the same basic lines of the original RCP problem. The world of agents, however, can be a lot more complicated. For this reason a generalization of RCP is a desired step for a more profound analysis of the protocol.

We obtain a generalization of RCP simply by loosening the rules connected to the RCP problem. For example the number or relations of players may be different or, more importantly, the card distribution can change and we can have cards that repeat themselves. It is also possible that the agents have only partial information available or none at all. Therefore generalized RCP cannot be solved with standard means. Questions could be a possible way of solving these generalized RCP problems. In order to do this, we need to learn about the role of questions in the archetypal RCP first.

3 Questions

We will use the framework developed in (Peliš & Majer, 2011). This framework is characterized by using set of answers methodology (a question is identified by a set of direct answers) and by an epistemic approach to questions. A question is understood as an epistemic statement, in which the inquirer informs the audience about her ignorance.

The language of erotetic epistemic logic is generated by the following BNF:

$$\psi ::= p \mid \neg\psi \mid (\psi \wedge \psi) \mid K_i\psi \mid ?_i\{\psi, \ldots\}$$

As in every normal modal logic we can define a dual operator $\hat{K}_i \psi \equiv \neg K_i \neg \psi$ with the meaning 'the agent i admits ψ'. A question is given by a set of direct answers $?_i\{\alpha_1, \alpha_2, \ldots \alpha_n\}$. In other words, an agent i asks : "Is it the case that α_1 or is it the case that $\alpha_2, \ldots?$"[4] RCP allows only for public updates and therefore we assume that also questions are public. Our scenario presupposes that some agent in the audience eventually chooses one of these options and announces the correct answer. These questions should satisfy three basic conditions: (1) answers are syntactically distinct, (2) there are at least two direct answers, and (3) the set of direct answers dQ^i is finite. In short, answers should be distinguishable and if there is only one option to answer, then the agent does not need to ask a question.

The central semantic notion is the one of *askability* of a question. This replaces the notion of truth in the semantics of standard propositional languages as it makes sense to ask a question in a certain situation, but it never makes sense to say that a question is true. A question $?_i\{\alpha_1, \alpha_2, \ldots \alpha_n\}$ is askable (for an agent i at a given state s of a model M)[5] iff it satisfies the following conditions:

1. $(M, s) \not\models K_i \alpha$, for each $\alpha \in dQ^i$

2. $(M, s) \models \hat{K}_i \alpha$, for each $\alpha \in dQ^i$

3. $(M, s) \models K_i(\bigvee_{\alpha \in dQ^i} \alpha)$

The first condition is called *non-triviality*—it is not reasonable to ask a question if the agent knows the answer. The second one, *admissibility*, requires that each direct answer is considered (by an agent i) as possible. And the third condition, *context*, means that at least one of the direct answers must be the right one (with respect to the knowledge of an agent i).

The third condition is problematic in some cases. It is superfluous for so called *safe questions*. For example, Anne's question "Does Bill hold 4?" is safe because the context of the question is given by tautology $(4_b \vee \neg 4_b)$.

Now, consider a question "Which natural number is greater than 13?", we can expect various answers to it, e.g., sets of some numbers greater than

---

[4] Representing a question as a set of direct answers does not correspond to the usual grammatical form of questions. A question can be more complex than just giving a list of direct answers. However, as our agents are simpletons, we can assume agents actually present the questions in this form.

[5] Our models are standard S5-models as it is usual in logics representing knowledge.

13, just one number greater than 13 or all the numbers greater than 13. Asking such a question an agent should only specify the form of expected answers (for example, she requires a singleton, some examples or a complete list), but she obviously does not list all possible answers.

Listing all the possible options in a real world situation could amount to a very long list. We could instead require that an agent lists a set of possible answers *she* has in mind and she is aware of the fact, that her list is incomplete. A new answer can be added by other agents and the agent then updates her knowledge. We allow this by omitting the context condition from the definition of askability.

This leads us to a generalization of the term *askability*. A *generally askable question* $?_i\{\alpha_1, \alpha_2, \ldots\}$ in a state s of a model M for an agent i is a question satisfying only the first two conditions of askability (non-triviality and admissibility).

4 Questionable RCP

Let us now combine RCP and questions. If we approach RCP with questions, we can get a system very similar to the original RCP. Now instead of announcing, Anne asks Bill, what cards he has. Does there exist an (RCP) *good question* that would allow Anne to learn about Bill's cards without Crow learning also?

First of all, Anne can ask Crow. We assumed all the agents are truth speaking and hence it is a valid option. This option is trivial, as Anne can ask single card questions (e.g., "Do you hold 4?") until she finds what card Crow has. The only necessary point is, Anne should ask also about her own cards. If she only asks about cards that she does not have, this protocol would not be safe. If done so, Crow cannot infer from Anne's questions what cards she has. However, in generalized RCP problems, Crow could have multiple cards. Depleting all the possibilities in that case could take a very long time.

Another important feature comes to light already at this early stage. Questions differ from announcements in that they do not allow for triviality. Hence Anne could not ask Crow about her cards in this fashion because she already knows that Crow does not hold her cards. The violation is clearly visible in this case. However, we need to keep it in mind later, in the more complex situations.

Let us return to Crow as the passive eavesdropper. A straightforward way how to introduce questions is to change the good announcement of Anne directly into a question. We demonstrate these options on the simple case of the archetypal seven card RCP.

Example 2 Anne's *good announcement* would be $[K_a(012_a \lor 034_a \lor 056_a \lor 135_a \lor 246_a)]$, hence the question would be $?_a\{K_b(012_a \lor 034_a \lor 056_a \lor 135_a \lor 246_a), \neg K_b(012_a \lor 034_a \lor 056_a \lor 135_a \lor 246_a)\}$. In other words, Anne asks Bill: "Do you think I hold one of these triplets?" Bill simply answers yes or no.

Here comes the twist. If he answers yes, Anne knows Bill knows her cards. However, because the question is based on the good announcement, Bill would never answer no. Therefore the question is trivial for Anne and it is unaskable according to the conditions. Let us try to construct a question that is askable for Anne. Anne could try to ask Bill about his cards instead of asking him about hers. Let us show the whole communication this time.

Example 3 Anne has her 012, Bill his 345, and Crow the 6. If Anne asks Bill over an open channel "Do you have one of these **025, 034, 056, 123, 145, 236, 346, 456**?" Bill will respond with a simple "no". At this moment Anne knows that Bill does not have neither 346 nor 456 (the remaining combinations contain always a card from Anne). Bill has no idea what is going on. Crow eliminates combinations that have his card and the remaining combinations he knows Bill does NOT have. Crow still has to take into account 15 options. Anne has only two left, 345 and 356. Hence Crow has according to Anne either 4 or 6. So the follow-up question to Bill would be "Are your cards from these 012345?". Now Bill says "yes". Thus Anne knows Bill's cards.

Is the first question askable for Anne? Yes, it is in the form $?_a\{(025_b \lor 034_b \lor 056_b \lor 123_b \lor 145_b \lor 236_b \lor 346_b \lor 456_b), \neg(025_b \lor 034_b \lor 056_b \lor 123_b \lor 145_b \lor 236_b \lor 346_b \lor 456_b)\}$. Both answers are a genuine possibility for Anne so the question is not trivial. Bill can also know Anne's cards, because if the protocol is fixed, he knows that a yes answer for the second question means that the set of cards was composed only of Anne's and Bill's cards.

We can see that even if the first question would be different, the outcome would be the same.

Example 4 Anne asks Bill "Do you have one of these **025, 034, 056, 123, 145, 236, 345, 456**?" Bill's answer is now positive. Hence for Anne, Bill

Erotetic Epistemic Logic in Private Communication Protocol 231

either has 345 or 456, the decisive cards being 3 and 6. She can make a similar second question as in the previous example. In the case of Crow this option presents a better outcome, as it has now five options to choose from.

Notice that the union of Crow's options covers the whole range of cards and that there is no intersection among the card sets. This is a property necessary for good announcements. However, it is *not* a necessity in the case of a good question. Natural language questions benefit from the possible incompleteness or even unintentional falsity of presented information. They also do suffer from this as we might need to construct a follow-up question.

If the second question is not as lucky, we can have a problem. The union of cards that are possible for Crow shows 012356. Crow learns that Anne does not know about 4. Hence she has 123 or 025.

Anne's original announcement can contain more options. This would worsen Crow's situation. Anne can construct her question based on the cards she takes as possible for Bill's hand, in our case 345, 346, 356, 456. She takes two of these and then she populates the length of the announcement (in the exemplar case up to eight cards) with random card combinations with approximately the same number of each card (in our case a card should be in the combinations three times on average).

Definition 1 (Good question) A good question *is a question from a cooperating agent that after being answered by another cooperating agent does not allow the eavesdroppers to know any of the cards of the cooperating agents' hand and it allows a cooperating agent to reduce the amount of possible combinations in another cooperating agent's hand to at least a half of the original amount of options (rounded down).*

As there are results characterizing for good announcements in RCP, it would be suitable to profit from them. For this purpose let us construct a good question from a good announcement. The algorithm for Anne to achieve this goal is the following:

- Prepare a good announcement
- Add half of Bill's possible hands (rounded up)
- Add possible hands containing always at least one card from Anne to even out card occurrences or change cards in present hands, following:
    - avoid duplicity

- do not alter Bill's possible hands
- maintain in all the hands at least one card from Anne, unless they are possible hands for Bill
- do not add too many options

The last point, about adding too many options, will be discussed later. For now, let us concentrate on the RCP with three agents as before, with a general card distribution $a \mid b \mid c$. We start with a case, when both cooperating agents have the same number of cards.

Proposition 1 *It holds in an RCP with a symmetrical card distribution between the cooperating agents, that if there is a good announcement, then there is a good question.*

Proof. Let us have a good announcement. If Anne adds to the good announcement half of the hands she takes as possible for Bill and then adds possible hands to the question to have all the cards represented in almost the same number. This is a good question. Based on Bill's answer, Anne can eliminate at least half of the options for Bill (rounded down). Because of the equality of card numbers, Crow has multiple options for Anne's cards and Bill's cards. □

Example 5 Let's assume a $4 \mid 4 \mid 1$ case, where Anne has 0123, Bill 4567 and Crow 8. Anne has the following good announcement: "I have some of these: **0123**, **0146**, **0158**, **0167**, **0246**, **0346**, **0368**, **0378**, **1248**, **1345**, 1356, 1457, 2357, 2457, 2478, 2678, 3458, 3567". Anne adds to this good announcement the possible cards for Bill's hand, for example 4567, 4678, 5678. The cards with the most occurrences are now 4 and 6, while 2 is with the least. Hence Anne does the following replacements: 1356 with 1235 and 0246 with 0236. Then she announces the question: "Do you have any of these: **0123**, **0146**, **0158**, **0167**, **0236**, **0346**, **0368**, **0378**, **1248**, **1345**, **1235**, 1457, 2357, 2457, 2478, 2678, 3458, 3567, 4567, 4678, 5678?"

Let us generalize to an RCP with a different number of cards but fulfilling the RCP conditions for card distribution from (Albert et al., 2005). We have two possible situations. Either Anne has more cards than Bill or Bill has more cards than Anne. The first case can be in the following form:

Example 6 For a 6 | 2 | 1 case, the smallest with Anne having more cards that allows good announcements according to Albert et al. (2005), the original good announcement of Anne can be: 012345, 012356, 012347, 012348, 014567, 135678, and so on. For the question Anne takes the pairs of options for Bills cards, that is 67, 68, 78 and takes two of them. Then she takes her good announcement and from each set takes two cards, at least one being her card. For example from 012345 she gets 01 or 135678 makes 37.

The second case can end up like this:

Example 7 For the distribution 3 | 4 | 1, a good announcement of Anne is 012, 034, 057, 136, 145, 235, 267. Anne adds to each of these an additional card from those she does not hold thus creating possibly the following question 0124, 0157, 0346, 1367, 1456, 2356, 3456, 3567 base is created, then she adds the possible cards of Bill, for example 3457, 3567, 4567 and finishes with the usual addition of new cards based on the quantity of the cards.

Proposition 2 *It holds in an RCP with an asymmetrical card distribution between the cooperating agents, that if there is a good announcement, then there is a good question.*

Proof. We need to distinguish the two cases, when Anne has more cards than Bill or vice versa.

In the first case, Anne takes at least half of Bill's possible hands. She adds hands created from each of hands in the good announcement. Each such hand has to contain at least one card from Anne's hand. Thereafter she adds more hands to keep the occurrence of individual cards even.

This is a good question. Anne added the possible cards of Bill. If she added Bill's actual hand, then she successfully eliminated half of the possible options. If she did not list the actual hand, she still eliminated half of the options. Because of the equal representation of cards, Crow has still enough possible hands and hence does not know which hand is the correct one.

In the second case, Anne adds possible Bill's cards to her good announcements and then adds at least half of the possible Bill's hands to the question. In the end, Anne has to compensate for the number of cards by adding additional hands. This again will be a good question. Bill can identify his hand if it is present or not. Anne, thanks to the construction of her question, eliminates half of the options from possible Bill's cards. Thanks to the equal distribution of cards, Crow remains again with multiple options. □

Corollary 1 *In RCP, if the card distribution has a good announcement, we can create a good question.*

The opposite does not hold and therefore good questions and good announcements are not equivalent. We can see this on the following example.

Example 8 The distribution 012 | 34 | 5 allows Anne to ask a good question. This can be done by asking Bill about the cards 02, 03, 12, 14, 25, 35, 45.

The feature that distinguishes a good announcement from a good question is that the proposition in the question can be false, but that answer is still informative. In the example, Bill will answer no. Thanks to that Anne knows Bills cards and hence can announce Crows card. A good announcement in this case would fail on the lack of cards. When Anne holds 012 and announces the options for her hand, she has little options before introducing triplets with two cards from her hand. Due to the limited amount of options, Anne will be able to present cards that contain the 5, hence get eliminated by Crow (any ending with 5), or repeat Bill's cards (i.e. 034, 134, 234). She cannot use two cards from her hand because she risks announcing a triple Bill would take as a possible hand for Anne (015, 025, 125).

5 Complexity

We promised to address also the problem of the amount of hands in a good question. Anne needs enough cards to make the question confusing for Crow and have a possibility to ask follow-up questions if necessary.

Observation 1 *Let us have an RCP problem with a distribution $a \mid b \mid c$. Then the upper bound for the number of hands for a good question is*

$$\binom{a+b+c}{b} \cdot \frac{1}{b-1} \qquad (2)$$

The idea is to take all the combinations of cards for Bill's hand size and limit the amount in proportion to that hand size. The limitation is necessary, because the larger the amount of cards held by Bill is, the larger the amount of possibilities will be. However, for a good question we do not need to list all of them. Obviously, the other constrains from the transformation algorithm still apply. Hence we try to have the same occurrence of cards and the question also contains at least half of the possible hands of Bill.

Erotetic Epistemic Logic in Private Communication Protocol

Because we are able to construct good questions from good announcements, we can use the bounds set for the size of good announcements in (Albert et al., 2005). We can alter these bounds by adding at least half of the number of cards based on Bill's possible hands. This follows straight from the way we constructed our good questions.

Observation 2 *Let us have an RCP problem with a distribution $a \mid b \mid c$. Then the lower bound for the number of hands for a good question is*

$$\frac{(a+b)(a+b+c)}{b(b+c)} + \left\lceil \frac{1}{2}\binom{b+c}{b} \right\rceil \tag{3}$$

Albert et al. (2005) presented two upper bounds. The first equation is a bound for the case when $b + c \leq a$, the second one is for all the other cases of good announcements.

$$\frac{(a+b+c)!(c+1)!}{(b+c)!(c+a+1)!} \left\lfloor \frac{a+c+1}{c+1} \right\rfloor \tag{4}$$

$$\frac{(a+b+c)!(c+1)!}{a!(b+2c+1)!} \left\lfloor \frac{(b+2c+1)}{(c+1)} \right\rfloor \tag{5}$$

We can construct our upper bounds based on these two equations.

Observation 3 *Let us have an RCP problem with a distribution $a \mid b \mid c$, where $b + c \leq a$. Then the suitable upper bound for the number of hands for a good question is*

$$\frac{(a+b+c)!(c+1)!}{(b+c)!(c+a+1)!} \left\lfloor \frac{a+c+1}{c+1} \right\rfloor + \left\lceil \frac{1}{2}\binom{b+c}{b} \right\rceil \tag{6}$$

Observation 4 *Let us have an RCP problem with a distribution $a \mid b \mid c$, where $b + c \not\leq a$. Then the lower bound for the number of hands for a good question is*

$$\frac{(a+b+c)!(c+1)!}{a!(b+2c+1)!} \left\lfloor \frac{(b+2c+1)}{(c+1)} \right\rfloor + \left\lceil \frac{1}{2}\binom{b+c}{b} \right\rceil \tag{7}$$

A proof that this works in the case of good announcement based questions is trivial. An interesting observation is that these boundaries work also for cases without good announcements.

Example 9 The distribution 012 | 34 | 5 has an upper boundary for the questions given based on the equation (6) and the calculation gives us 7 as answer. The lower bound is 6. The example used the good question of length 7, namely 02, 03, 12, 14, 25, 35, 45. A shorter version can be 02, 03, 12, 14, 35, 45.

The protocol how to make a good question without using a good announcement is very simple and quite similar to the original protocol.

- Take randomly half of the possible hands of Bill (rounded up)
- Add hands composed of at least one card from Anne's hand while
  - not repeating any hand
  - maintaining an approximately equal occurrence of cards in the question
  - do not exceed the number of cards given by the upper bound

The fact that we have this protocol does not mean we can construct a good question for any card distribution (for example 1 | 1 | 1 is still an unsolvable case). However, it does allow us to address a larger number of card distributions than the good announcements did. At least for this reason, it is worth thinking about questions in the RCP.

6 Conclusion

We explored the possibility of using questions instead of announcements in the protocol of the Russian Cards Problem. We concentrated on the RCP class with two cooperating agents and one eavesdropper without repetition of cards. We showed that questions provide more general solutions for this class: for any solution in the form of a good announcements there is a solution in the form of a good question, but not vice versa. We presented an example for which a good question exists, but a good announcement does not. We also gave some complexity bounds for the amount of hands necessary for a good questions.

We plan to use questions in more general Russian cards problems which include more players and allow for card repetitions. This might require extension of the framework of questions and employing weaker background epistemic systems than from S5. A useful step to elaborate on the topic is also the introduction of trust among agents. The basic inspiration comes from the article (Baltag & Smets, 2009). There general plausibility frames are endowed with 'Radical' Upgrade and 'Conservative' Upgrade. This would allow the agents to have a hierarchy similar to the contemporary hierarchy of trust protocol servers.

References

Albert, M., Aldred, R., Atkinson, M., van Ditmarsch, H., & Handley, C. (2005). Safe Communication for Card Players by Combinatorial Designs for Two-step Protocols. *Australasian Journal of Combinatorics, 33*, 33–46.

Baltag, A., & Smets, S. (2009). Talking Your Way into Agreement: Belief Merge by Persuasive Communication. In *Proceedings of the Second Multi-Agent Logics, Languages, and Organisations*. Aachen: Federated Workshops.

Duan, Z., & Yang, C. (2009). Generalized Russian Cards Problem. In D.-Z. Du, X. Hu, & P. Pardalos (Eds.), *Combinatorial Optimization and Applications* (pp. 85–97). Berlin: Springer.

Peliš, M., & Majer, O. (2011). Logic of Questions and Public Announcements. In *Eighth International Tbilisi Symposium on Logic, Language and Computation 2009, Lecture Notes in Computer Science* (pp. 145–157). Berlin: Springer.

van Ditmarsch, H. (2003). The Russian Cards Problem: A Case Study in Cryptography with Public Announcements. *Studia Logica, 75*, 1–32.

van Ditmarsch, H., van der Hoek, W., van der Meyden, R., & al. (2006). Model Checking Russian Cards. *Electronic Notes in Theoretical Computer Science, 149*, 105–123.

Ondrej Majer
Institute of Philosophy, Czech Academy of Sciences
The Czech Republic
E-mail: majer@flu.cas.cz

Michal Peliš
Charles University in Prague
Institute of Philosophy, Czech Academy of Sciences
The Czech Republic
E-mail: michal.pelis@ff.cuni.cz

Petr Švarný
Charles University in Prague
The Czech Republic
E-mail: svarnypetr@gmail.com

The Role of Perspectives in the Interpretation of Defeasible Reasoning

GIACOMO TURBANTI

Abstract: Non-monotonicity in logic is a symptom that may have many causes. In the formalisation of defeasible reasoning, an epistemic diagnosis has largely prevailed according to which some inferences are non-monotonic because they are provisionally drawn in the absence of relevant or complete information. The Gabbay-Makinson rules for *cumulative consequence relations* are a paradigmatic example of this epistemic approach. In this paper a different approach to defeasible reasoning is introduced, based on the idea of inferential perspectives. According to this approach, some inferences are non-monotonic because they are drawn as from another reasoner's perspective. Rules are introduced and discussed for a *Perspectival Calculus*, which show both similarities and interesting differences with respect to *cumulative systems*.

Keywords: perspectival calculus, inferential perspectives, defeasible reasoning, non-monotonic Logic, substructural logics.

1 Introduction

According to Makinson (2005), the essence of non-monotonic reasoning is shown by Sherlock Holmes' deductions. Surely, none of these is, strictly speaking, a deductive inference. Rather, they represent the familiar fallible practice of drawing the most reasonable inferences from the defective knowledge bases that we are usually provided with. In such practice, when new information forces us to withdraw some conclusion, still we recognise it as the most reasonable one to be drawn in the previous circumstances. So, in this sense, non-monotonic reasoning is construed as dealing with epistemic uncertainty and lack of information. My aim in this paper is to dig deeper into this interpretation. I suggest that the sort of uncertainty involved in non-monotonic reasoning is to be construed as due to perspectiveness. In fact, I suggest to construe non-monotonic reasoning as the sort of reasoning we perform when we infer *as if* from another point of view. Discrepancies between different points of view engender fallibility. Surely epistemic discrepancies are a paradigmatic example, but the rationale of the approach may be generalised to moral perspectives, aesthetic perspectives, etc.

Now, generally speaking, a perspective is a point of view on something. To look at something in perspective is to look at it from *a certain* point of view, as opposed to another possible point of view. The very notion of a perspective implies the idea of a multiplicity of perspectives. Considering multiple perspectives is something we constantly do, no matter how selfish or egocentric we are. I don't want to explain how it happens that we *can* do that, I just assume that we do. My aim is rather to analyse some of the features of this sort of reasoning. So, for instance, while writing this paper it is in my best interest to try to put myself in the reader's shoes in order to be as clear and compelling as possible. Notice, however, that while I try to reason *as if* I were the reader, I do not really drop my own perspective. Instead, I try to figure out how the reader would reason *from* his perspective on the view which I take to be correct as far as the topic of this paper is concerned. Such a projection would likely result in a partial representation of what I take to be the fact of the matter, so that reader's reasoning as interpreted by me will be defeasible. In this sense, perspectival reasoning engenders an asymmetric relation between the interpreter's own point of view and the interpreted one. Let me pick two evocative labels for these points of view, and let me call them respectively the "slave" perspective and the "master" perspective, where, in the typically Hegelian fashion, the slave perspective is the favoured one. Overall, my basic question here could be put like this: what is it to reason from the *master* perspective?

The paper is structured as follows. In Section 2, Gabbay's approach to non-monotonic reasoning will be presented and the analysis of cumulative systems will be offered as a paradigmatic case of such approach. In Section 3, the semantic framework of the perspectival approach will be introduced. In Section 4, the new approach will be exploited in order to single out the main features of the proof system for a *Perspectival Calculus*. Eventually, in Section 5, a small comparison is drawn between the two systems.

2 Non-monotonic logic

In logic, *Monotony* is a property of consequence relations. Given a deductive system, $\mathbf{S} = \langle \mathfrak{Fm}_L, \vdash_S \rangle$, where $\mathfrak{Fm}_L$ is the algebra of formulae over the logical language L and a countably infinite set of propositional variables, and the relation $\vdash_\mathbf{S} \subseteq P(Fm_L) \times P(Fm_L)$ is a consequence relation for $\mathbf{S}$.

A closure operator is customarily associated with consequence relations s.t. $\gamma_{\vdash_\mathbf{S}}(X) = \{\phi \in Fm_L \mid X \vdash_\mathbf{S} \phi\}$. In this sense, a classical consequence operator γ satisfies the following conditions, for $A, B \subseteq Fm_L$:

(Inclusion) $A \subseteq \gamma(A)$
(Idempotence) $\gamma(A) = \gamma(\gamma(A))$
(Monotony) $A \subseteq B$ implies $\gamma(A) \subseteq \gamma(B)$

Hence, **S** is said to be monotonic if the consequence operator $\gamma_{\vdash_\mathbf{S}}$ satisfies the condition of *Monotony*. It's worth recalling here that the characterisation of logical consequence is completed by the requirement of *invariance under substitution*. Thus, let a substitution σ be a complete endomorphism on $\mathfrak{Fm}_L$, so that, for $A \subseteq Fm_L$:

(Structurality) $\sigma(\gamma(A)) = \gamma(\sigma(A))$

Now, in this framework, it seems that the obvious way to define non-monotonic logic is to drop, or at least to weaken, the condition of *Monotony*. But, as a matter of fact, there is something more to say about this. Let's begin by rephrasing the above conditions in terms of the (collectively) equivalent ones for *consequence relations*.

(C1)   if $a \in X$ then $X \vdash a$
(C2)   if $Y \vdash a$ for all $a \in X$, and $X \vdash b$, then $Y \vdash b$
(C3)   if $X \vdash a$ and $X \subseteq Y$, then $Y \vdash a$
(C4)   if $X \vdash a$ then, for every substitution σ, $\sigma[X] \vdash \sigma(a)$

As it's easy to see, (C3) is redundant for it is implied by (C1) and (C2). The fact that *Monotony* is implied by *Inclusion* and *Transitivity* is actually the first problem to solve in order to get off the ground with non-monotonic logic. It seems that a non-monotonic deductive system must drop, or weaken, either (C1) or (C2). So, for instance, on the one side, Relevance Logic and Linear Logic may be interpreted as adopting a notion of deducibility according to which $X_1, \ldots, X_n \vdash a$ is valid if $X_1, \ldots, X_n$ are *intensionally* and *all* used in the deduction of a. Gabbay (1985) picked the other option by weakening *Transitivity*. Following this lead, he proposed the two weakened conditions of *Cumulative Transitivity* (CT) and *Cautious Monotony* (CM), such that, for $A, B \in Fm_L$:

(CT) $A \subseteq B \subseteq \gamma(A)$ implies $\gamma(B) \subseteq \gamma(A)$
(CM) $A \subseteq B \subseteq \gamma(A)$ implies $\gamma(A) \subseteq \gamma(B)$

In the following I will generally refer to the logics defined according to this strategy by the somehow inaccurate yet convenient name "Gabbay's logics". As a consequence of the failure of full *Monotony*, it is characteristic of this approach to define systems which are *weaker* than classical logic in the sense that for any $A \in Fm_L$, $\gamma_{\models}(A) \subseteq \gamma_{\vdash_s}(A)$, where $\models$ indicates the classical monotonic consequence relation. Makinson named this property, *Supraclassicality*. It is important to notice straight away that, as established in (Makinson, 2005, Theorem 1.1), no non-trivial supraclassical consequence relation is substitution invariant, and that, therefore, Gabbay's logics fail *Structurality*.

2.1 The cumulative system C

Although Gabbay's systems constitute a variegated collection of non-monotonic logics with different properties, in this paper I'll focus just on the cumulative system **C**. Indeed, it is a paradigmatic example because it satisfies all the properties originally identified by Gabbay. Kraus, Lehmann, and Magidor (1990) provide a Gentzen-style presentation of **C**, here reported in Table 1.

$$(\text{Id}) \; \frac{}{\alpha \mathrel{|\!\sim} \alpha}$$

$$(\text{CT}) \; \frac{\alpha \mathrel{|\!\sim} \beta \quad \alpha \wedge \beta \mathrel{|\!\sim} \gamma}{\alpha \mathrel{|\!\sim} \gamma} \qquad (\text{CM}) \; \frac{\alpha \mathrel{|\!\sim} \beta \quad \alpha \mathrel{|\!\sim} \gamma}{\alpha \wedge \beta \mathrel{|\!\sim} \gamma}$$

$$(\text{LLE}) \; \frac{\alpha =\!\!|\!\models \beta \quad \alpha \mathrel{|\!\sim} \gamma}{\beta \mathrel{|\!\sim} \gamma} \qquad (\text{RW}) \; \frac{\alpha \mathrel{|\!\sim} \beta \quad \beta \models \gamma}{\alpha \mathrel{|\!\sim} \gamma}$$

Table 1: The system **C**

Some remarks are in order. First, consider *Left Logical Equivalence* (LLE) and *Right Weakening* (RW). As we already noticed, one of the crucial features of Gabbay's systems is that their consequence relations are construed as extensions of the classical one. In order to manage the interaction between the non-monotonic and the classical consequence relations, the Gentzen-style presentation of the preferential system involves these two rules with side-conditions. So, for instance, *Supraclassicality* is entailed by (Id) and (RW). Second, notice that the presentation doesn't involve structural rules. However it is claimed that the language of **C** is the classical one, so it is implied that *some* standard structural and operational rules are valid. In fact, it is also claimed that some rules are "derived" in **C**, as, for instance, the rule for the introduction of conjunction in the consequences:

$$\text{(And)} \frac{\phi \mathrel{\vdash\!\sim} \psi \quad \phi \mathrel{\vdash\!\sim} \chi}{\phi \mathrel{\vdash\!\sim} \psi \wedge \chi}$$

In the same sense, other rules establishing the classical behaviour of logical operators, like *Permutation* and *Contraction*, may be considered as derived rules. Yet, there are exceptions. Obviously, **C** must fail *Thinning:*

$$\text{(WL)} \frac{\phi \mathrel{\vdash\!\sim} \chi}{\psi \wedge \phi \mathrel{\vdash\!\sim} \chi}$$

Less obviously, it must also fail plain *Cut*, the "easy part" of the *Deduction Theorem* and *Contraposition*:

$$\text{(Cut)} \frac{\phi \mathrel{\vdash\!\sim} \psi \quad \psi \mathrel{\vdash\!\sim} \chi}{\phi \mathrel{\vdash\!\sim} \chi} \quad \text{(EHD)} \frac{\psi \mathrel{\vdash\!\sim} \phi \to \chi}{\phi \wedge \psi \mathrel{\vdash\!\sim} \chi} \quad \text{(CP)} \frac{\phi \mathrel{\vdash\!\sim} \psi}{\neg \psi \mathrel{\vdash\!\sim} \neg \phi}$$

The reason is that, if any of these is added to **C**, it would reestablish monotonicity in the form of *Thinning*.

2.2 Preferential semantics

Preferential semantics is a solid achievement in the study of Gabbay's logics. Originally introduced by Shoham (1988), it essentially became the gold standard thanks to Makinson (1988) and Kraus et al. (1990).

Definition 1 *Let L be a propositional language and V a set of valuations, then a* cumulative model M *is a triple* $\langle S, l, \prec \rangle$ *where S is an arbitrary set, whose elements are called 'states'; $l: S \longmapsto V$ is a 'labelling function' which assigns a valuation to each state;* $\prec$ *is an asymmetric relation on S.*

The basic purpose of preferential semantics is to allow the selection of models which are *minimal* in the order of the reasoner's preferences, i.e. the preferred ones.

Definition 2 *Let* $M = \langle S, l, \prec \rangle$ *be a preferential model, and* $X \subseteq S$, *then* $s \in X$ *is minimal in* X *iff* $\forall t \in X, t \not\prec s$.

In order to avoid infinite descending chains in preferential orders, sets of states are required to be *smooth* (or *stoppered*).

Definition 3 *Let $M = \langle S, l, \prec \rangle$ be a preferential model, then M is smooth iff for all $X \subseteq S$, $\forall t \in X$, either t is minimal in X or $\exists s \in X$ s.t. $s \prec t$ and s is minimal in X.*

A cumulative consequence relation is then defined using these minimal states.

Definition 4 *Let $M = \langle S, l, \prec \rangle$ be a cumulative model and $\widehat{\phi} = \{s \in S \mid l(s)(\phi) = 1\}$, then for any $T \subseteq L$, $\phi \in L$, $T \mathrel{\vdash\mkern-10mu\sim}_M \phi$ iff for all $s \in S$ minimal in $\widehat{T}$, $l(s)(\phi) = 1$.*

Notice that cumulative models are closed under finite intersection, in the sense that $\widehat{(\phi \wedge \psi)} = \widehat{\phi} \cap \widehat{\psi}$. This guarantees the validity of (AND). Notice also that the *smoothness* condition is essential for the validity of (CM): suppose M were not smooth, then $\alpha \wedge \beta \mathrel{\vdash\mkern-10mu\sim}_M \gamma$ might be false while both $\alpha \mathrel{\vdash\mkern-10mu\sim}_M \beta$ and $\alpha \mathrel{\vdash\mkern-10mu\sim}_M \gamma$ are true just because $\widehat{\alpha}$ has no minimal element.

Thus we have the following soundness and representation theorems

Theorem 1 (Kraus et al., 1990) *For any cumulative model M, all the rules of* **C** *are satisfied by $\mathrel{\vdash\mkern-10mu\sim}_M$.*

Theorem 2 (Kraus et al., 1990) *Let $\mathrel{\vdash\mkern-10mu\sim}$ be any supraclassical cumulative consequence relation, then there is a smooth cumulative model M such that $\mathrel{\vdash\mkern-10mu\sim}_M = \mathrel{\vdash\mkern-10mu\sim}$.*

3 The perspectival approach

Now, the harsh list of results in Section 2 may engender some discomfort with respect to a certain lack of interpretative unity of the overall picture offered by Gabbay's approach. *Why* supraclassical logics are not structural? *Why* just *Deduction Theorem* and *Contraposition* fail? Indeed, on the one side, there are proofs, but somehow they do not seem to be explanatory enough. On the other side, there is semantics, but it seems quite fragmented in apparently unrelated conditions required for different rules, which convey a certain feeling of *adhoc*-ness.[1] Not even the epistemic interpretation seems to come to one's aid here, its only contribution amounting to the suggestion that all these failures represent different ways in which the reasoner is ignorant and thus uncertain. But what sort of ignorance is responsible for the failure of *Structurality*, for instance? The discomfort is especially

---

[1] A comprehensive survey of the semantics for Gabbay's non-monotonic logics may be found in (Gabbay & Schlechta, 2010).

Perspectives in the Interpretation of Defeasible Reasoning 245

striking if this picture is compared with the unitary analysis of some other non-monotonic logics, like Relevance Logic and Linear Logic, in terms of *substructural* logics. Thus, the purpose of the rest of this section will be to extend such analysis to Gabbay's non-monotonic logics for defeasible reasoning.

3.1 Substructural logics

A logic is said to be *substructural* if its Gentzen system lacks any of the structural rules.[2] A Gentzen system is a logic **G** whose consequence relation $\vdash_\mathbf{G}$ is defined on a set Seq of sequents, thus $\vdash_\mathbf{G} \subseteq P(Seq) \times P(Seq)$. A sequent over a logical language L is a pair $\langle \Gamma, \Delta \rangle$, indicated here as $\Gamma \rhd \Delta$ (unless differently required), where Γ and Δ are finite sequences of formulae in L, so that a sequent $\phi_1, \ldots, \phi_n \rhd \psi_1, \ldots, \psi_m$ has *type* (n, m). A sequent system may be reformulated as a 2-dimensional system—i.e. a system $\mathbf{G} = \langle \mathfrak{Fm}, \vdash_\mathbf{G} \rangle$, where $\vdash_\mathbf{G}$ is defined on Fm^2—if structural operators are used to combine in a single formula sequences occurring on the left-hand side and on the right-hand side of a sequent. Notice also that, in this sense, the properties of the operators chosen to represent the combination of formulae play a crucial role. In fact, clear correspondences can be stated between structural rules and algebraic properties of the combinators. Here, obviously *Thinning*, *Weakening* and *Cut* have to be put under close observation. The relevant correspondences are recalled in Table 2, which is adapted from (Dunn, 1993). $*, +$ are generic meet and join operators and $\phi \rhd \psi$ is a generic sequent.

Let me first recall some basic notions in algebraic semantics. A logical matrix is an ordered pair $\langle \mathfrak{A}, F \rangle$, where $\mathfrak{A}$ is an algebra and $F \subseteq A$ represents the set of designated values. An interpretation is an homomorphism $h: Fm \longmapsto A$. So, consider again the set Fm_L of the formulae of a given logical language L: a logical matrix $\langle \mathfrak{A}, F \rangle$ is a *model* of a logic $\mathbf{S} = \langle \mathfrak{Fm}_L, \vdash_\mathbf{S} \rangle$ if and only if for every homomorphism h and every $\Gamma \cup \{\phi\} \in Fm_L$, if $h(\gamma) \in F$ for all γ in Γ, and $\Gamma \vdash_S \phi$, then $h(\phi) \in F$. In this sense, F is said to be a *deductive filter* of S. Thus, logical matrices can be used to define semantic interpretations of deductive systems.

Then, consider the *Lindenbaum algebra* of the logic S. We define an equivalence relation $\equiv_\mathbf{S}$ on formulae as coderivability, which means that $\equiv_\mathbf{S} = \{\langle \phi, \psi \rangle \mid \phi \vdash_\mathbf{S} \psi \text{ and } \psi \vdash_\mathbf{S} \phi\}$. It's easy to check that $\equiv_\mathbf{S}$ is also

---

[2]An extensive presentation of substructural logics and their semantics can be found in (Paoli, 2002).

	Logic	Algebra
(WL)	$\dfrac{\Gamma \triangleright \phi}{\Gamma, \psi \triangleright \phi}$	Lower Bound: $a * b \leq a$
(WR)	$\dfrac{\phi \triangleright \Delta}{\phi \triangleright \psi, \Delta}$	Upper Bound: $a \leq a + b$
(Cut)	$\dfrac{\Gamma \triangleright \phi \quad \Delta, \phi \triangleright \psi}{\Gamma, \Delta \triangleright \psi}$	Isotonicity: $\begin{array}{l} a \leq b \Rightarrow a * c \leq b * c \\ a \leq b \Rightarrow c * a \leq c * b \end{array}$

Table 2: Logic-Algebra correspondences

a congruence relation on $\mathfrak{Fm}$, so the quotient algebra $\mathfrak{Fm}_L/\equiv_S$ can be defined. Let $[\phi]$ be the equivalence class containing ϕ. A partial order is defined s.t. $[\phi] \leq [\psi]$ iff $\phi \vdash_S \psi$. Thus, eventually, the partially ordered groupoid $\langle Fm/\equiv_S; \leq, *\rangle$ can be considered, where $[\phi] * [\psi] = [\phi, \psi]$.

Now, we are ready to establish the connections between algebraic properties and logical rules. On the one side, suppose that $*$ satisfies Lower Bound. So, suppose $\gamma \vdash_S \phi$, then $[\gamma] \leq [\phi]$, and so $[\gamma] * [\psi] \leq [\phi]$ by Lower Bound. Hence $\gamma, \psi \vdash_S \phi$, and we have *Thinning*. On the other side, suppose that $*$ satisfies Isotonicity. So, suppose $\gamma \vdash_S \phi$ and $\delta, \phi \vdash_S \psi$, then $[\gamma] \leq [\phi]$ and $[\delta] * [\phi] \leq [\psi]$. But then $[\delta] * [\gamma] \leq [\delta] * [\phi]$ by Isotonicity, and so $[\delta] * [\gamma] \leq [\psi]$ by transitivity of $\leq$. Hence $\delta, \gamma \vdash_S \psi$, and we have *Cut*.

3.2 Introducing inferential perspectives

Our goal is to show how non-monotonic reasoning originates from the interaction of different inferential perspectives. So, let's consider two inferential perspectives as represented by two consequence relations defined in terms of sequents over different languages $L_1 = \{\wedge, \vee, \rightarrow, \neg\}$ and $L_2 = \{\curlywedge, \curlyvee, \Rightarrow, \neg\}$, namely $\vDash_1 \subseteq P(Seq_1) \times P(Seq_1)$ and $\vDash_2 \subseteq P(Seq_2) \times P(Seq_2)$. In order to mark the difference between the two sets of sequents we'll indicate sequents $(\phi, \psi) \in Seq_1$ as $\phi \blacktriangleright \psi$ and sequents $(\phi, \psi) \in Seq_2$ as $\phi \triangleright \psi$. What it is crucial to notice is that sequents from different sets will have different structural operators: $\bullet$ and $\circ$ respectively. In fact, the two different inferential perspectives involve two different ways of *taking contents together*. Thus, $\phi_1 \bullet \ldots \bullet \phi_n \blacktriangleright \psi_1 \bullet \ldots \bullet \psi_m$ and $\phi_1 \circ \ldots \circ \phi_n \triangleright \psi_1 \circ \ldots \circ \psi_m$ are interpreted respectively as $\phi_1 \wedge ... \wedge \phi_n \to \psi_1 \vee ... \vee \psi_m$ and $\phi_1 \barwedge ... \barwedge \phi_n \Rightarrow \psi_1 \veebar ... \veebar \psi_m$. The next step is to define the rules for each. At this point the standard strategy in Gabbay's logics is to look at the rules for classical consequence relation and weaken them enough to block *Thinning*. I now suggest instead to stick to standard rules for both our consequence relations and look at their interaction. So first, let's suppose that both $\models_1$ and $\models_2$ satisfy all Gentzen's structural rules for a classical propositional sequent calculus. Then we have to introduce the asymmetry envisaged in Section 1, and we do it in terms of the following property of perspective *projection*:

(Projection) $\quad \phi \blacktriangleright \psi$ implies $\phi \triangleright \psi$

Projection introduces an asymmetric relation to the extent that the one inferential perspective, $\models_2$, can just directly 'represent' the other, $\models_1$, but not *viceversa*.

3.3 Perspectival lattices

We now define a suitable framework for algebraic semantics. So, let's consider the logic $\mathbf{SL} = \langle \mathfrak{Fm}_{L_2}, \models_{\mathbf{SL}} \rangle$ for the *slave* perspective, where, as before, $\models_2 \subseteq P(Fm_{L_2}) \times P(Fm_{L_2})$ and $\phi \models_{\mathbf{SL}} \psi$ iff $\models_2 \phi \triangleright \psi$. Then consider the congruence relation $\equiv_{\mathbf{SL}} = \{\langle \phi, \psi \rangle \mid \phi \models_{\mathbf{SL}} \psi \text{ and } \psi \models_{\mathbf{SL}} \phi\}$ on Fm_{L_2} and let $\mathfrak{SL} = \langle Fm_{L_2}/\equiv_{\mathbf{SL}}; \sqcup\!\!\!\sqcup, \sqcap\!\!\!\sqcap, \mathbf{0}, \mathbf{1}\rangle$ be the Lindenbaum algebra for $\mathbf{SL}$, where $[\phi] \sqcup\!\!\!\sqcup [\psi] = [\phi \barwedge \psi]$, $[\phi] \sqcap\!\!\!\sqcap [\psi] = [\phi \veebar \psi]$ and $[\phi] \sqsubseteq [\psi]$ iff $\phi \models_{\mathbf{SL}} \psi$. In fact, it can be shown that $\mathfrak{SL}$ is a Boolean algebra.

Now, the *master* inferential perspective must be introduced. We could try to proceed as before by considering the logic $\mathbf{MA} = \langle \mathfrak{Fm}_{L_1}, \models_{\mathbf{MA}} \rangle$, where $\phi \models_{\mathbf{MA}} \psi$ iff $\models_1 \phi \blacktriangleright \psi$, the congruence relation

$$\equiv_{\mathbf{MA}} = \{\langle \phi, \psi \rangle \mid \phi \models_{\mathbf{MA}} \psi \text{ and } \psi \models_{\mathbf{MA}} \phi\}$$

and the Lindenbaum algebra

$$\mathfrak{MA} = \langle Fm_{L_1}/\equiv_{\mathbf{MA}}; \sqcap, \sqcup, \mathbf{0}, \mathbf{1}\rangle,$$

where $[\phi] \sqcap [\psi] = [\phi \wedge \psi]$, $[\phi] \sqcup [\psi] = [\phi \vee \psi]$ and $[\phi] \sqsubseteq [\psi]$ iff $\phi \models_{\mathbf{MA}} \psi$. $\mathfrak{MA}$ will be a Boolean algebra as well.

However this is a wrong way for us to proceed: what is obtained is a second algebra, while we required a second inferential perspective in the same algebra. In fact, we can't first construct Lindenbaum algebras for the two inferential perspectives, because they would have nothing in common. The right idea is pictured in Figure 1: in part (a) straight lines represent a

Hasse diagram for the ordering relation ⊑ and dotted lines represent a Hasse diagram for the relation ⊑ and $Fm_{L_2} \cap Fm_{L_1} = \{a, b, \mathbf{0}, \mathbf{1}\}$; in part (b) the desired structure is represented as resulting from the interaction of the two algebras.

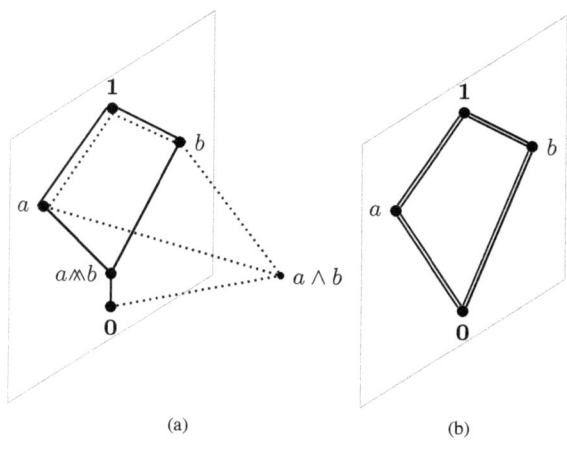

Figure 1

Therefore, we introduce the second inferential perspective by means of two further binary operations $\otimes, \oplus$ on $\mathfrak{Fm}_{L_2}$ as the greatest lower bound and least upper bound in $Fm_{L_1} \cap Fm_{L_2}$ with respect to both the "slave" order ⊑ and the "master" order ⊑. Crucially, we let $\otimes, \oplus$ be *partial* operations, so that $\mathfrak{M}\mathfrak{A} = \langle Fm_{L_2}; \otimes, \oplus, \mathbf{0}, \mathbf{1} \rangle$ is a bounded *weak partial lattice*, as defined in (Grätzer, 2011, Definition 78). Then we define a second ordering relation $\preceq$ on $\mathfrak{Fm}_{L_2}$ as follows: let $\phi \preceq \psi$ iff $\phi \otimes \psi \equiv \phi$ iff $\phi \oplus \psi \equiv \psi$, whenever $\phi \otimes \psi$ and $\phi \oplus \psi$ exist. In the following, the existential requirement will not be mentioned where obvious. Notice that there will be members of $\mathfrak{Fm}_{L_2}$ which are not ordered by the $\preceq$ relation. This is why in (Grätzer, 2011, Theorem 84) the stricter notion of *partial lattice* is introduced. However that is not really a problem for us, for *all* members of $\mathfrak{Fm}_{L_2}$ are ordered by the ⊑ relation. Moreover, in compliance with *Projection*, we don't allow $\otimes$ and $\oplus$ to be defined at will, but we impose $\phi \preceq \psi \Rightarrow \phi \sqsubseteq \psi$. Let me thus put forward a definition for these complex algebraic structures.

Perspectives in the Interpretation of Defeasible Reasoning

Definition 5 *A perspectival lattice is an algebra* $\mathfrak{PL} = \langle PL; \sqcup, \sqcap, \otimes, \oplus \rangle$, *where* $\mathfrak{SL}_{\mathfrak{PL}} = \langle PL; \sqcup, \sqcap \rangle$ *is a lattice, with lattice order* $\sqsubseteq$; $\mathfrak{MA}_{\mathfrak{PL}} = \langle PL; \otimes, \oplus \rangle$ *is a weak partial lattice, with lattice order* $\preceq$; *for any* $a, b \in PL$, *if* $a \preceq b$ *then* $a \otimes b = a \sqcap b$, $a \oplus b = a \sqcup b$.

The following definition provides a strategy for the construction of perspectival lattices.

Definition 6 *Let us have two bounded lattices* $\mathfrak{SL} = \langle S; \wedge_S, \vee_S, 0_S, 1_S \rangle$ *and* $\mathfrak{MA} = \langle M; \wedge_M, \vee_M, 0_M, 1_M \rangle$. *Then* $\mathfrak{SL} \sqcap \mathfrak{MA} = \langle S; \preceq \rangle$ *where for any* $a, b \in SM$, $a \preceq b$ *iff* $a \leq_S b$ *and* $a \leq_M b$.

Lemma 1 *Let* $\mathfrak{SL}$ *and* $\mathfrak{MA}$ *be as in Definition 6, then* $\mathfrak{SL} \sqcap \mathfrak{MA}$ *is a weak partial lattice.*

Proof. It is routine to check the properties of the definition of weak partial lattice. □

We will require the two basic lattices to share the bounds. This is a plausible requirement, for it amounts just to require an agreement at least on what is *truth* and what is *falsity*. It is hard to see how two rational agents could build any common ground in their inferential perspectives in the absence of such a minimal requirement. In fact, as we will see in a moment, this requirement corresponds directly to the requirement of *Smoothness* for cumulative models in Definition 3 above.

4 Reasoning in perspective

In this Section the rules will be considered for a *Perspectival Calculus*, representing reasoning from the *master* perspective. We want both our consequence relations to satisfy all Gentzen's structural rules for a classical propositional sequent calculus. We now proceed to analyse what happens to these rules in the interaction of the two perspectives.

The *Cut* rule is crucial for the integration of perspectives since it should establish how inferences drawn from different perspectives can be concatenated. Pictorially speaking, to concatenate two sequents is to 'cut away' the *same* content on the right of the first one and on the left of the second one so to create a slot where the two sequents can 'hang' together. As we've seen, two algebraic properties are required to perform the operation: *Isotonicity*, which guarantees that the formulae to be cut really have the same content even when combined with other ones, and *Transitivity*, which authorises the concatenation of the sequents. In our model $\mathfrak{PL}$, $\otimes$ is isotonic with respect to $\preceq$, $\sqcup$ is isotonic with respect to $\sqsubseteq$, and both $\preceq$ and $\sqsubseteq$ are transitive. In fact non-perspectival *Cut* rules are valid.

$$(\text{Cut.1}) \frac{\phi \triangleright \psi \quad \gamma_\circ \psi \triangleright \chi}{\gamma_\circ \phi \triangleright \chi} \qquad (\text{Cut.2}) \frac{\phi \blacktriangleright \psi \quad \gamma_\bullet \psi \blacktriangleright \chi}{\gamma_\bullet \phi \blacktriangleright \chi}$$

However, neither *Isotonicity* nor *Transitivity* apply 'inter-perspectively'. In fact inferences can't be directly concatenated when different perspectives are considered, because different contents are obtained by combining formulae with different operators in different inferential perspectives. Therefore the following cases fail.

$$(\text{Cut.3}) \frac{\phi \blacktriangleright \psi \quad \gamma_\circ \psi \blacktriangleright \chi}{\gamma_\circ \phi \blacktriangleright \chi} \qquad (\text{Cut.4}) \frac{\phi \triangleright \psi \quad \gamma_\bullet \psi \triangleright \chi}{\gamma_\bullet \phi \triangleright \chi}$$

In fact, even if ϕ entails ψ, nothing guarantees that the content of ψ *concatenated with γ in the slave perspective* can be inferred in the master perspective from the content of ϕ *concatenated with γ in the slave perspective*. This failure with *Cut* was expected after Section 2. But it is also expected that a slight modification of (Cut.3) and (Cut.4) would provide us with the required weakened rules for *Cut* in perspective.

$$(\text{Cut.5}) \frac{\phi \blacktriangleright \psi \quad \gamma_\circ \phi_\bullet \psi \blacktriangleright \chi}{\gamma_\circ \phi \blacktriangleright \chi} \qquad (\text{Cut.6}) \frac{\phi \triangleright \psi \quad \gamma_\bullet \phi_\circ \psi \triangleright \chi}{\gamma_\bullet \phi \triangleright \chi}$$

These are barely *Cut* rules. Here the inference from ϕ to ψ directly guarantees that the content of ϕ *concatenated with γ* is the very same the content of 'ϕ and ψ' *concatenated with γ*. In particular, by *Projection* we have the following rule which corresponds to (CT) in system **C**.

$$(\text{CT}) \frac{\phi \blacktriangleright \psi \quad \gamma_\circ \phi_\circ \psi \blacktriangleright \chi}{\gamma_\circ \phi \blacktriangleright \chi}$$

Let's now focus on *Thinning*. This rule is the distinguishing feature of monotonic deductive systems, for it establishes that the validity of an inference is not compromised by the addition of new contents to its premises. As we've seen, in algebraic models this rule is connected to the property of fusions being lower bounds. Since, in our model $\mathfrak{PL}$, $\otimes$ and $\sqcup$ satisfy the Lower Bound property with relation respectively to $\preceq$ and $\sqsubseteq$, again, non-perspectival rules (WL.1) and (WL.2) are valid.

$$\text{(WL.1)} \frac{\phi \triangleright \chi}{\psi_\circ \phi \triangleright \chi} \qquad \text{(WL.2)} \frac{\phi \blacktriangleright \chi}{\psi_\bullet \phi \blacktriangleright \chi}$$

The interesting cases involve the interaction of the two perspectives. However, one of these cases, (WL.3) follows just by *Projection*. The converse case is more problematic. In fact, since, in general ⊔ doesn't satisfy Lower Bound with relation to $\preceq$, (WL.4) will fail.

$$\text{(WL.3)} \frac{\phi \triangleright \chi}{\psi_\bullet \phi \triangleright \chi} \qquad \text{(WL.4)} \frac{\phi \blacktriangleright \chi}{\psi_\circ \phi \blacktriangleright \chi}$$

Yet, here again, a slight modification of the rule would do. What is required is for *Thinning* to be accepted only when it is safe enough, i.e. when the content of ϕ and $\psi_\circ \phi$ is the same. That can be obtained by imposing $\phi \blacktriangleright \psi$ as a further premise. The modified rule corresponds to Gabbay's proposal for the weakening of *Monotony*, which is part of the definition of system **C** as (CM).

$$\text{(CM)} \frac{\phi \blacktriangleright \chi \quad \phi \blacktriangleright \psi}{\psi_\circ \phi \blacktriangleright \chi}$$

The most interesting part of the latter proof concerns the role of the premise $\phi \blacktriangleright \psi$. *Projection* guarantees that the way ϕ and ψ are taken together in the two inferential perspectives is the same. That allows to apply *Thinning* in the interaction of the two perspectives.[3]

In the new framework we may also try to expand our horizons a little. So, it's easy to realise that the case of *Weakening* on the right is the dual of the case of *Thinning* on the left. While relying on the previous analysis, we speed up the discussion here. Again, non perspectival cases are trivially valid. *Projection* guarantees the case of weakening with $\bullet$ with respect to $\triangleright$, but the converse case fails. Yet again a slight modification would do. What is required here is that the content of χ and $\chi_\circ \psi$ is the same, and that can be done by imposing $\psi \blacktriangleright \chi$. Let me identify the modified rule as *Cautious Weakening*.

---

[3] A remark is in order about the validity of *Cautious Monotony*. The lattice-theoretical framework adopted here obscures the fact that (CM) may fail in algebraic structures which are not closed under lower bounds: if $\phi \otimes \psi$ didn't exist the rule would fail. The existence of the minimum element thus corresponds to the condition of *Smoothness*. However, recall that such a guarantee is the result of a requirement we imposed in Section 3.

$$(\text{CW}) \frac{\phi \blacktriangleright \chi \quad \psi \blacktriangleright \chi}{\phi \blacktriangleright \chi \circ \psi}$$

Notice how (CW) quite naturally follows from the formal approach adopted here.[4] Indeed, when the interaction of different inferential perspectives is put into focus, such a rule has also a certain intuitive appeal. Suppose in fact you are trying to draw inferences in my perspective, and suppose we are both willing to infer χ from ϕ. Now you can certainly draw the further conclusion ψ *or* χ. Why wouldn't I do the same? As a matter of fact I would be willing to draw ψ *or* χ *in my perspective*. The point is obviously that ψ *or* χ *in your perspective* might not have the same content of ψ *or* χ *in my perspective*.[5]

5 Conclusion

Let the *Perspectival Calculus* **PerC** be the system defined by (CT), (CM) and (CW) of Section 4. **PerC** is a non-monotonic logic. It is defined on the basis of the integration of different inferential perspectives. It is important to acknowledge that it shows important differences with respect to Gabbay's non-monotonic logics. The best way to appreciate these differences is in the light of *Supraclassicality*. Recall that *Supraclassicality* is the property of a consequence relation $\vdash_S$ to be weaker than classical consequence relation $\models$ in the sense that for any $\phi \in L$, $\{\psi \in L \mid \phi \models \psi\} \subseteq \{\psi \in L \mid \phi \vdash_S \psi\}$. Therefore *Supraclassicality* is not really a rule of any logical system. It is important to bear in mind this obvious fact while approaching the issue in this framework, for it might be tricky to distinguish it from *Projection*. Here the fact that Kraus et al. (1990) use the same notation both for sequents and for consequence relations doesn't help, and one might easily be misled by abuses of notation. So, the question here is if the consequence relation defined in terms of the rules of **PerC** is supraclassical. It's easy to realise that it is not. In fact, even if both the inferential perspectives we are considering

---

[4]Notice also that (CW) may fail in algebraic structures which are not closed under upper bounds. So, as in the dual case of (CM), it is essential that the basic lattices of the algebraic semantics share the bounds.

[5]It's worth wondering why this rule has never crossed anyone's mind in the study of Gabbay's logics. The most relevant reason is probably the simplest one: (CW) has nothing to do with monotony. Or so it seems. As a matter of fact, the study of non-monotonic logics is focused on what happens on the left of the turnstile, so that the treatment of disjunction is often overlooked. Nonetheless it has proved to be quite crucial in the representation of different properties of preferential models, as shown for instance in (Freund, 1993). More recently, in (Gabbay & Schlechta, 2008), it is noticed that the very *Cumulativity* may fail in non-monotonic logics whose models are not closed under finite unions.

are classical, classical rules may fail in their interaction. So, in particular, the validity of (LLE) and (RW) in perspectival reasoning depends on how side conditions are interpreted in **PerC**: if $\phi \models \psi$ is construed as a rule in the classical master perspective then (RW) will be valid, while if it is construed as a rule in the classical slave perspective then (RW) will fail. In particular (RW) will fail whenever the side condition involves connectives from the slave language $L_2 = \{\wedge, \vee, \Rightarrow, \neg\}$. This is why the rules (And) and (EHD) fail in **PerC**. So, overall, *Perspectival Reasoning* essentially differs from *Cumulative Reasoning*. Nonetheless, I hope I have shown that the analysis of the interaction between different inferential perspectives provides an interesting framework for the interpretation of defeasible reasoning.

References

Dunn, J. M. (1993). Partial Gaggles Applied to Logics with Restricted Structural Rules. In P. Schroeder-Heister & K. Došen (Eds.), *Substructural Logics* (pp. 63–108). Oxford: Claredon Press.

Freund, M. (1993). Injective Models and Disjunctive Relations. *Journal of Logic and Computation, 3*, 231–247.

Gabbay, D. M. (1985). Theoretical Foundations for Nonmonotonic Reasoning in Expert Systems. In K. R. Apt (Ed.), *Logics and Models of Concurrent Systems* (pp. 439–457). Berlin: Springer.

Gabbay, D. M., & Schlechta, K. (2008). Cumulativity without Closure of the Domain under Finite Unions. *Review of Symbolic Logic, 1*, 372–392.

Gabbay, D. M., & Schlechta, K. (2010). *Logical Tools for Handling Change in Agent-based Systems*. Heidelberg: Springer.

Grätzer, G. (2011). *Lattice Theory: Foundation*. Basel: Springer.

Kraus, S., Lehmann, D. J., & Magidor, M. (1990). Nonmonotonic Reasoning, Preferential Models and Cumulative Logics. *Artificial Intelligence, 44*, 167–207.

Makinson, D. (1988). General Theory of Cumulative Inference. In M. Reinfrank, J. de Kleer, M. L. Ginsberg, & E. Sandewall (Eds.), *Nonmonotonic Reasoning* (pp. 1–18). Heidelberg: Springer.

Makinson, D. (2005). *Bridges from Classical to Nonmonotonic Logic*. London: College Publications.

Paoli, F. (2002). *Substructural Logics: A Primer*. Dordrecht: Kluwer.

Shoham, Y. (1988). *Reasoning about Change*. Cambridge, MA: MIT Press.

Giacomo Turbanti
University of Pisa
Italy
E-mail: turbanti.giacomo@gmail.com

www.ingramcontent.com/pod-product-compliance
Lightning Source LLC
Chambersburg PA
CBHW051042160426
43193CB00010B/1042